INTRODUCTION TO ECOLOGY AND ENVIRONMENT

Encyclopaedia of Environmental Pollution and Awareness in 21st Century Series

INTRODUCTION TO ECOLOGY AND ENVIRONMENT

Edited by
V.K. Prabhakar

ANMOL PUBLICATIONS PVT. LTD.
NEW DELHI - 110 002 (INDIA)

ANMOL PUBLICATIONS PVT. LTD.
4374/4B, Ansari Road, Daryaganj
New Delhi - 110 002

Introduction to Ecology and Environment

First Edition, 2001

ISBN 81-261-0910-6

PRINTED IN INDIA

Published by J.L. Kumar for Anmol Publications Pvt. Ltd., New Delhi - 110 002 and Printed at Mehra Offset Press, Delhi.

Contents

PREFACE

Today, the cry of 'pollution' is heard from all the nooks and corners of the world, and pollution has become a major threat to the very existence of mankind on this earth. It is the major challenge of our times.

The pollution of various resources has gone to such an extent that we are unable to breathe fresh air and drink fresh water. On one hand the advancements of science and technology have added to the human comforts by giving us automobiles, electrical appliances, supersonic jets, space crafts, better medicines, better chemicals to control harmful insects and other pests, etc., but, on the other hand, they have given us a very serious problem to face that is pollution. Pollution can be defined as an undesirable change in the physical, chemical, or biological characteristics of the air, water, or land that can harmfully affect health, survival, or activities of humans or other living organisms.

Pollution does not have to cause physical harm, but it may merely interfere with human activities. Chemicals spewed into the air or water from an industrial plant may be harmful to humans and other organisms living near the plant. If expensive pollution controls are required, the plant may be forced to shut down.

India faces serious problems of environmental pollution. Water pollution from untreated community and industrial wastes flowing into our rivers and streams, increasing use of fertilizers and pesticides, dumping of organic wastes etc., are issues that need to be tackled on the priority. Air pollution, long regarded as the bane of industrialised nations looms large over our country.

Here, we are presenting this *Encyclopaedia of Environmental Pollution and Awareness in Twenty-first Century* in fifty volumes with a view to make people conscious about the seriousness of pollution in the present time and its remedies.

Editor

1 FLORAL AND FAUNAL DISTRIBUTION IN INDIA

The unique biogeographic position of the Indian subcontinent adds further interest to its wild life. The region lies at the confluence of Ethiopian, Palaearctic and Indo-Malayan faunas and possesses a number of interesting components from each of these realms as well as several peculiarly indigenous genera. The Ethopian element is represented by the chinkara, the hyena and the ratel, the Palaearctic by lynx, wolf, markhor, urial and hangul, the Chinese by red panda and musk deer and the Indo-Malayan by the hoolock gibbon, the elephant, the goat-antelopes and the mouse-deer. The genera restricted to India include Melursus (sloth bear), Antilope (blackbuck), *Tetracerus* (chousinga or fourhorned antelope) and Boselaphus (nilgai).

The presence or absence of an animal or plant in a certain region is determined by ecological and historical factors. Populations of organisms can become established in a region only if the range of conditions under which the species or individual can thrive (its ecological balance) is consistent with the sum of the conditions prevailing there. Animals and plants, then are living indicators of the characteristics of their environment, their ranges mark the places in which environmental conditions are the same or similar. At the same time, the evolution and distribution of species casts light upon the geological evolution of various parts of the earth, and upon the course of global changes in climate and vegetation.

Ecological Sub-Division

The present physiography and climate of India together with the great faunal changes that took place during the late pliocence period have imparted to its flora and fauna, a wide variety of habitats, each supporting its own distinctive type of flora and fauna. It is convenient, when looking at the relationships between the animals, to recognise three subregions

(1) *The Himalayan mountain system:* The fauna in the higher reaches of this region has strong affinities with the palaearctic. Some of the animals like bears, wild pigs, tigers and snow leopards are found on both sides of the Himalaya in the palaearctic and oriental regions. They seem to have come to India through migrations along the narrow passage way presented by the flood plains and the gorges of the Brahmaputra Ganges river systems or the early stages of formation of the Himalayan massif, which at that time, did not form the unsurpassable barrier as it does today. There are altitudinal as well as East-West variations in this region. Three district subzones, each with its characteristic assemblage of animals are now recognised. The first covers the Himalayan foothills from the eastern frontiers of Kashmir to Assam. The second covers the higher altitudes in the Western Himalayas from Kashmir including Ladakh to Kumaon and the third the Eastern Himalayan sub-region.

(2) *The Peninsular-Indian sub-region:* It was once linked by continuous woodland savanna with the Ethiopian region of North Africa. Desert barriers now separate the two areas but the faunas retain certain similarities, sharing such animals as the lion, cheetah, leopard, hyaena, jackal and antelopes.

(3) *The Tropical evergreen forest of Indo-Malayan Sub-region*: It has retained its natural rainforest vegetation since the Pliocence. It has similarities both with the Indian as well as Malayan and Indo — Chinese fauna. The rain-forest region of Western Ghats is also linked with its through climatic similarities and past tropical forest associations. This is rich and highly complex environment and contains a wealth of species of flora and fauna. Species such as the red panda, gibbon, treeshrew, tapir, giant, squirrel and flying lemur are confined almost exclusively to this region.

The Himalayan Mountain System

The Himalayan massif is a comparatively recent phenomenon on the geological time scale and in the past was not a barrier to animal movements that it is today. Consequently, much of the high altitude fauna of this area extends over the whole complex of mountain ranges and is typical of both the oriental and palaearctic regions.

Altitude however, influences the distribution of flora as well as fauna in this region. It is estimated that every 1000mt. rise in altitude is equivalent to a 10° increase in latitude. Thus near arctic conditions can be found at elevations of 3500 mt. in the Himalayan ranges.

In addition to the influence of altitudinal change on ecology, there are marked east-west variations due to decrease in monsoon rainfall towards the west. The western Himalaya is a region of low rainfall, snowfall is heavy, and conditions are more akin to temperate regions. The eastern Himalaya have high rainfall, snowfall is confined to very high altitudes and at lower altitudes conditions resemble the tropical rain forests. The temperate region in the eastern Himalaya starts above 2745mt. as against 1500mt. in the West. The flora differs both in physiognomy and structure and the fauna is also very different having affinities with the Chinese fauna of the palaearctic.

The Himalayan Foothills

The Himalayan foothills are charactrised by bhaber & tarai tract consists of thick boulder deposits where the water table is very low. The conditions are relatively driver here. The tarai region consists of thick deposits of fine silt, the drainage is poor and the water table is high. The natural monsoon forest extends well intothis region. The dominant species is sal. In its natural state, the tarai is characterised by all grassy meadows with savanna vegetation. This is one of the richest areas for the typical big mammals of Northern India. The elephant, the sambar, the swamp deer, cheetal, hog deer, barking deer, wildboar, all abound in this area. This is also the famous tiger country immortalised in Jim Corbett. Panther and wild dogs are the copredators and the hyena, the jackal, the camp followers of the tiger, scavenging of the remains of the tiger kill. Both the black and the sloth bears are found in this area. Amongst the rodents, procupine is the most prominent animal. The great Indian One horned rhinoceros, which once ranged all along the Himalayan foothills is now confined to a few pockets in Assam and Nepal tarai and is considered a rare species. Another animal which shares this habitar with the rhinoceros and which, together, with the gaur, has the distinction of being the largest bovid in the world, is the wild buffalo. The only other habitat where this great bovid now exists is the Godavari catchment of the States of Madhya Pradesh, Orissa (where it is now reported to be extinct) and Andhra Pradesh. Wild buffalo is considered a threatened species.

The brow-antlered deer is another rare deer found in this region confined to Keibul Lamjo Sanctuary in Manipur. This deer whose number in the wild is estimated at less than 50 now, lives on a floating mass of vegetation deposits called 'Phumdi' and due to its floating habitar, has been named the 'dancing deer'of

Manipur by the late E.P.Gee. This deer has the doubtful distinction of probably being the most threatened deer in the world.

Most of this area, especially the tarai belt has however been colonised by agriculture and much of the wildlife has disappeared from here. The large elephant herds which used to migrate all along the foothills have been fragmented by agriculture, construction of reservoirs for hydroelectric generation, and other human artifacts. The tiger, probably the most famous animal of India, has become rare with the disappearance of a large part of its habitat, and with it, its prey base of large ungulates. In some forest pockets where it exists surrounded by agricultural fields as in the Kheri District, it is directly in conflict with man, posing problems of man-eating and cattle-lifting. It is however being specially protected by a total ban on its killing, and implementation of special 'Project Tiger'. Two of the eleven project areas are located at 'Corbett Park' and Manas Sanctuary' in this belt. It is now one of the most vulnerable zones for its wildlife.

This area also supports some animals found nowhere else in the orientals region. The gangetic gharial, the sole representative of the family gavialidae, occurs in the Himalayan tributaries of the river Ganges. Until very recently, it was thought that hispid hare and the pigmy hog had become extinct. It was only 1971 that these two species were re-discovered in Manas Sanctuary in Assam. The fauna of this region, specially in the eastern parts, is also not very well documented. As recently as in 1950s, again in the Manas Sanctuary, E.P.Gee was able to establish the existence of a new species of monkey, the golden langurs which is now named after him as Presbytis geei. These langurs are of golden chestnut colour in winter and paler during the hot season; the young are parctically white. They exist in some localised patches on either side of the Manas river in India as well as in Bhutan.

The High Altitude Region of Western Himalayas

This region consists of the belt of coniferouspine forests occupying the altitudinal zone from 1500 mt. to 2500mt. The Rhododendron, dwarf hill bamboo and birch forests mixed wtih alpine pastures extend above the pine belt upto the snow line; the cold desert plateau of Ladakh exists in the extreme North West. These higher reaches above the coniferous belt represent the alpine zone. The pine belt mostly acts as a transition zone for the rich fauna of the alpine zone in the higher altitudes. During winter, when the environment in the alpine zone becomes extremely cold, a number of species move down in this region. With the summer thraw, the alpine pastures again turn green and become the grazing grounds of most of these animals.

One of the typical animals of the high altitude cold desert regions of Ladakh and Tibet is the wild ass which is found in the palaearctic deserts of Asia. Though identified as a rare and endangered species, it is still fairly common in Rupshu, Changthang and Chang Chenmo areas of Ladakh in the Indian region. The bovid family is well represented in this part of the Himalaya, particularly in the higher altitudes. More species of wild goats and sheep live in this region than anywhere else. The largest bovid, the yak, has been domesticated for centuries by the people of Tibet and Ladakh but wild yak can still be found. They are black with long black horns and a little white on the muzzle.

Three species of wild goats occur in the Indian Himalaya. The highest coniferous forests are roamed by thar, a typical goat with rather short horns set close together on top of the head. Thar are gregarious animals grazing in herds while sentinels mount a look out for possible danger. Even higher, on the rocky slopes and cliffs above the tree line, lives the markhor, the finest of all goats with its long horns shaped like thick, heavy, cork screws. The fur of markhor is rich reddish brown turning to grey in winter. Like all wild goats, it is an amazingly agile climber. In winter, markhor retreats to lower slopes where forage is more plentiful. They even climb along the branches of evergreen oaks to browse its leaves.

The third goat found in the Western Himalaya is the ibex. But unlike thar and markhor which are limited to the Himalaya, ibex has a much wider range, inhabiting mountainous regions all over Central Asia from the Himalaya to the Altai and extending westwards as far as Span. The ibex live mainly above the tree line grazing between rocky cracks and no meadows just below the snow line.

Amongst the sheep also, there are three species of wild sheep typical of the Western Himalaya. Nayan is the largest species of wild sheep and extends from Ladakh to Tibet. It constitutes of the 15 races or sub-species of Argali sheep which ranges from the Himalaya to Siberia. Marcopolo saw and described one of these in the 13th Century from which it was named as Macropolos sheep, renowned for its magnificient horns which curve round in a huge sweep, making more than a complete circle.

The bharal or blue sheep occurs in the western Himalaya extending to Sikkim and Nepal, it has characters intermediate between the sheep and the goats. Bharals are found at altitudes of 3600m to 5000m and feed on the rich and abundant grass on the alpine meadows. Bharals feed in herds during the summer months except for the old males which forage at higher elevations. During the winter of the year, the ewes congregate in small harems round the male. Their colour of slaty blue merges perfectly with the rocky outcrops of their habitat and is difficult to spot them if they are motionless.

The shapu or urial is the smallest of the wild sheep and inhabits grassy mountain slopes usually below the tree line. The urial has probably contributed to the stock of domesticated sheep.

Two kinds of antelopes, the chiru and the Tibetan gazellen complete list of Himalayan and tibetan bovids. Both graze on the hıgh cold plateau regions in Tibet and adjoining areas. The chiru has a swollen muzzle which may be an adaptation for beathing cold and rarefed air.

All these animals are adapted to cold and harsh conditions. Their coat is thick in the winter and thinner in the summer and the seasonal coats are often differently coloured so that they merge with the background and camouflage the animals. All these animals except the two antelopes, are sharp eyed and marvellously surefooted rock-climbers.

Two deer species are also characteristic of the Western Himalayan high alpine zone. The hangul or Kashmir stag is the local representative of the European red deer. This deer is mostly confined to the Dachigam sanctuary in Jammu and Kashmir near Srinagar. The shou or the Sikkim stag is another race of the red deer (much larger than the hangul) which occurs in the Chumbi valley in the Tibet and Sikkim. (Reported to be extremely rare now).

The musk deer is another deer which inhabits the high altitudes bordering the tree line. This deer is much valued for its musk pods which are collected from the males and used in the manufacture of perfumes. Due to large scale poaching, this deer has now become threatened throughout its range. A special project called the Musk-deer Breeding project has been recently launched by the Government of Uttar Pradesh.

Many smaller mammals live above the treeline. The typical among them are the marmots. The Himalayan or Bobak marmot is a stoutly built animal with a short bushy tail and very small ears.Marmots live in large colonies. They live in burrows and feed on grass around their burrows. In a autumn they drag dry grass into these burrows and hibernate in this warm bedding until the spring when the young are born. In the high mountains of Kashmir also lives a little known woolly flying squirrel, a cumbersome creature capable of gliding like its forest relatives.

Pikas or mouse hares live in the mountains at altitudes from 2500mt. to well above 4000mt. They are similar in appearance to guinea pigs but belong to the rabbit and hare family. These small mammals overcome the inhospitable climate by burrowing underground.

Though there are few species in the alpine zone, a food chain is still apparent. There are predators among birds and mammals. The Himalayan golden eagle preys on marmots, snow cocks and snow patridges nad the little Tibetan weasel ranges high to feed on birds and their eggs.

Among mammals, the most beautiful animal is the new leopard with a long thick coat of smoky grey with black rosettes. It is smaller than the average panther. It is mainly solitary, hunting at night. It preys on wild sheep and goat and the musk deer and follows them up and down the mountainous slopes on their seasonal migration. The snow leopard has been hunted for its beautiful coat which is most highly priced. It is now regarded as a threatened species. The other predators include the wolf, the fox, smaller cats like the Pallas cat and the black and the brown bears.

Amongst the birds in this region, mention must be made of a number of pheasants which characterise the Himalayan mountain ranges. Amongst these are the Himalayan monal pheasant, the western tragopan the satyr tragopan, the koklass, the whitecrested khalj and the cheer pheasant. Pheasants are birds with most attractive plumage. They are hunted for their meat as well as feathers and as a group are considered as rare and threatened.

Griffon vultures, lammergiers, choughs and ravens are the scavengers that follow man as well as wandering yak or wild sheep, wherever they roam.

The Eastern Himalayan Sub-Zone

From the dense evergreen and semi-evergreen vegetation of the foothills in the Eastern Himalaya, the character of vegetation changes at altitudes of 1525 m to 1830 m. Oaks, magnolias, laurels and birches covered with moss and ferns replase the sal, silk-cotton trees, and giant bamboos of the foothills. At about 2745 to 3660m one enters the coniferous forest of pine" fir, yew and junipers. There is an undergrowth of scrubby Rhododendrons and dwarf bamboos. The boundary between the temperatee zones and the tropical forests of lower altitudes and foothills is not sharply defined. Due to high humidity and much higher rainflall, lichens mosses, orchids and other epiphytes cover the tree trunks. The temperate belt occupies much higher altitudes than in the west. Snowfall is less common. The animal life in the temperate region is different from the western Himalaya and is characterised by the presence of Indo Chinese fauna. The red panda, hog badgers, ferret badgers, crestless procupines are typical species of this area.

Three kinds of goat antelopes also occur in the Eastern Himalaya and are relatives of the European chamois, Serow is a goat like animal with a black coat, large ears, white beard and a greyish mane. It lives mostly in the eastern Himalaya in forests around 2000 mts. and also extends into the tropics in Malaya and Sumatra where it occurs at lower altitudes on steep lime stone outcrops.

Goral is a smaller goat antelope found throughout the Himalaya living on rugged grassy hill sites and on rocky grounds near the conifer forests. It is remarkably agile on the precipitous cliffs. Goral has been a favourite game animal of the sportsmen visiting Himalayan hill resorts, like the chamois in Europe.

Takins are the largest of the goat antelopes and characterise the eastern Himalayan forests near the timber line. In summer, large herds graze above the timber line but in winter the animals break up into smaller groups and migrate to grassy valley lower down the mountains.

The Peninsular India Sub-Region

Peninsular India is the true home of the Indian fauna. It is characterised by the raised plateau land of the Deccan extending northwards into the flood plains of the Indo-Gangetic basin and westwards into the Great Thar desert of Rajasthan. The Thar or the Indian desert is the area lying on the west of the Aravalli ranges and just east of the Indus Valley. Ecologically, this may also include the salt flats of the Little Rann of Kutch which lie to the southwest of Thar desert. The rainfall in the desert areas is less than 250 mm per year and the ecology of the area is completely different from the rest of peninsular India.

The rest of Peninsular India is a tropical land mass stretching across the Tropic of Cancer from roughly 8 N latitude of nearly 20° N latitude. It is characterised by monsoon rains which are mainly confined to only 3 or 4 rainy months followed by long dry spells and high temperatures.

It is possible to identify two broad zones in this region. Firsly the tropical deciduous woodlands characterising peninsular India and its extension into the drainage basin of the Ganges river system. The second is the Desert region of Rajasthan or the Thar with connected salt flats of Little Rann of Kutch and the desert fringe.

Peninsular India and the Drainage Basin

In its natural state the region is the home of tropical moist deciduous to tropical dry deciduous and scrub vegetation depending upon the variation in rainfall and humidity. The northern and eastern extensions having relatively higher rainfall have sal as the predominant species while the southern plateau is characterised by teak as the principal species. The Western Ghats and the central belt lying to the west of it, is a region of very high rainfall and is characterised by evergreen vegetation, its flora and fauna being a kin to the evergreen rain forest of north-eastern India. In the drier north-western portions bordering the Rajasthan desert and Aravalli hills, the trees are more scattered and thronyscrub species predominate. The forest gives way to more open savannah habitat. These areas have also been subjected to recurring fires and grazing which have resulted in the artificial spread of savannah at the expense of tlhe monsoon forest. Originally this area was very rich in its wild ungulate fauna but due to adverse biotic factors it has now greatly suffered.

Water holes or perennial pools of water are the common feature of the deciduous scene. During a long dry summer in this region, most of the water in the small rivulets or nallas dries up except for a few pools, or springs. Some pools are also formed in natural depressions which shrink to small water holes surrounded by muddy apron in the summer. These water holes and other perennial sources of water play an extremely important role in the distribution of local animal life during the pinch period of summer, and one of the best ways of seeing animals during the summer is to set up an observation post by the side of a water hole.

The open and deciduous nature of the forest with a large number of trees being of low height, make conditions more favourable for herbivorous ungulates. The seasonal nature of food supply also favours larger ungulate species that can build up reserves to tide them over the long lean months. The area, therefore, has the potential of supporting a high ungulate biomass. It has been estimated for the Kanha Reserve in M.P. that at the height of the pinch period in summer it can support 60-70 kg. of wild ungulates plus 180kg. of domestic livestock per hectare.

Of the larger inhabitants of the Indian deciduous forests are animals that are more widely spread in the rainforest. Elephant, muntjak the unrifagous sambar deer and the ubiquitous wild boar all occur in both habitats. The gaur the huge characteristic bovid of Central India has its counterparts in rain forests of Malaya where it is called the seladang. India also has its own species of mouse deer closely related to those of South—East Asia.

In addition, however, Peninsular India has several species of deer and antelope that do not occur in the rain forests further east. Perhaps the most attractive of these is the cheetal or the Axis deer. This elegant spotted deer is usually found in herds of 5-10 but in the meadows of Kanha or the Chaurs of Corbett National Park, huge groups of several hundred may congregate. This deer is both a grazier and a browser and has a high breeding potential.

A smaller stockier relative of the Axis deer is the hog deer, but it prefers rather swampier meadows and is found more in the tarai belt of the Himalayan foothills. Another deer which shows preference for the moist river valley meadows is the swamp deer of barasinga. This is a large deer and the stag have fine antlers containing upto 12 points. A sub-species of this deer or the hardground barasinga is now confined only to the Kanha reserve in Central India. The barasinga is now an endangered species.

The sambar is the largest of all Indian deer and has a very wide range of distribution ranging from the throny scrub forest of eastern Rajasthan to semi-evergreen forest is eastern India and from the foothills of the Himalaya in the north to the tip of the Indian Peninsula in the south. The sambar lives in small herds, is mostly nocturnal and though urifagous is mainly a browser. It prefers thick shady glades in the deciduous forest. The muntjak or the barking deer is mostly a solitary deer and like sambar also likes thickly wooded country. The warning calls of the barking deer, or kakar as it is locally known, together with the characteristic 'ponking' call of the sambar when tiger is on the move are mong the more familiar sounds in the Indian jungles.

The deer drop their bony antlers regularly and replace them with new growth. The new antlers are first covered woth a soft velvet like covering which is rubbed off after the full growth of the antler has been achieved. Only the stags have antlers and the hinds are without any natlers or horns. The cycle of antler development is quite closely connected with the rutting behaviour of the males.

Antelopes and other bovids are the other group of ungulates present in the Indian Peninsula. Unlike deer, they have permanent horns. One woodland species of

Indian antelopes called fourhorned antelope has the unique distinction of having two large horns at the back. This animal lives in pairs inhabiting wooded hilly parts of country. Two large antelopes, the nilgai and the blackbuck also inhabit the open habitats in the deciduous woodlands but are more characteristic of the semi-desert and arid areas. The chinkara gazelle is the smallest of the antelopes. They live in small parties and prefer more open habitats.

The three main predators of the Indian woodlands are the Wild dog or dhole, the tiger and the leopard. The dhole also characterises the rainforest environment. It is an animal which hunts in packs and works in well organised hunting parties and once it surrounds is prey it is very difficult of even big ungulates like the sambar to escape. Even the tiger is scared of hunting packs of dhole. The Indian leopard belongs to the same species as those that inhabit the rainforests but they are generally spotted rather than the melanic or black panther form of south-east Asia. The leopard is normally a nocturnal hunter and its prey includes the wild pig, the smaller deer and monkeys. Often it may hunt from tree branches where it may sit in ambush.

Probably the most famous animal of the forests of Peninsular India as well as the Himalayan foothills is the tiger. It is also now India's 'National Animal'. Though believed to have its origin in the Palaearctic realm, the Indian tiger or the Royal Bengal tiger as it is more often called, has come to be recognised as the type species (Panthera tigris tigris) amongst the 8 races of the tiger that are recognised all over its range of distribution in the Palaearctic and the oriental regions. Its strength and courage, its beauty and phantom like grace made tiger hunting a sport of the princes over the ages. More recently it has been shot out by indiscriminate poachers hunting in jeeps with modern weapons and through hunting trips, organised by shikar agencies which charged heavy fees from visiting foreigners. The habitat of the tiger has also been shrinking very rapidly and with the corresponding decrease in its wild ungulate prey base, it has to take to cattle lifting and occasional man-eating, thus directly coming into conflict with man. This magnificient animal whose population was estimated to have been not less than 40,000 at the turn of the century was reduced to less than 2,000 animals in the early 1970s. A special conservation project 'Project tiger' to save this animal organised by the Govt. of India is in progress in 11 selected reserves in collaboration with World Wildlife Fund.

The other two big savannah cats of Peninsular India the cheetah and the lion also had their origin outside India in the Ethiopian region. The cheetah or the hunting leopard is the fastest moving land animal over short distances it can achieve speeds of upto 60 miles (100 km) per hour. The cheetah was once a common sight in this region, more particularly as a pet kept by the princes of India to hunt that fast moving antelope the blackbuck and enjoying the chase as a sport. Unfortunately the cheetah is now extinct in India. The last reliable mention of a live wild cheetah was made by K.M. Kirpatrick who reported seeing one on the road near Chandragiri in a low hill tract in the heard of South India on the night of 28-29 March, 1952 while he was out driving.

The Asiatic lion like its African cousins is an animal of the open savannah. It is slightly heavier in build than its African relatives. The young are spotted, the adults have bigger tail tufts and the males have shorter manes. Unlike the tiger which is a solitary animal, lions are social animals. They live in family group called 'prides'. Hunting is mainly done by the females. The lion once had a wide ranged of distribution throughout waestern and Central India across to Persia, Arabia and the Middle East. It has now disappeared from all its previous range and is now disappeared from all its previous range and is now confined to a small pocket of forests inthe Gir Sanctuary of the Gujarat State in Western India. Here it has been specially protected and their numbers, according to the latest counts carried out in may 1979, are a bare 205. A special project called 'The Gir Lion Sanctuary Project' is being implemented by the Govt. of Gujarat for the protection of this rare cat which once was the national animal of India and adorns the seal of the Government of India.

Two other carnivores of African origin are the striped hyena and the jackal. Both are more of scavengers, feeding off the remains of a tiger kill, waiting for the bigger predators to move away to feed on the remains. Hyenas are powerful animals with immensely strong jaw and molar teeth which enable them to crush the bones of even the largest carcass to eat the marrow. They also kill young and sick animals and have the reputation of digging up buried corpses unless covered with heavy stones. Occasionally, hyenas have also been accused of child lifting from urban areas, Jackals are also quite a familiar part of the rural scene of India where their howls at night are one of the common sounds one hears. They serve a useful social function, cleaning the debris of villages and small towns, but at the same time they can be a menace to poultry. Jackal has been regarded as a vermin among India's wildlife but recently it is noticed that jackal skins are being utilised by the fur trade in Kashmir on a very large scale which may pose a threat to this once abundant animal.

The elephant which has a wide distribution in the Himalayan tarai is missing from most of the Central

Indian plateau particularly the states of Madhya Pradesh, Andhra Pradesh and Maharashtra but reappears in Karnataka, Kerala and Tamil nadu and extends into Ceylon. It is difficult to find a suitable explanation for its absence from much of the Central Indian Plateau. In the South the elephant is associated with the famous Kheda operation in the State of Karnataka. The elephant Kheda was a well organised operation in which hundreds of trained elephants and thousands of men were utilised in capturing big herds of wild elephants by encircling them and driving them ito a huge stockade prepared for this purpose. The whole show could be watched through camouflaged grand-stands and has been described by some as one of the greatest shows on earth.

The biggest bovid of the Indian region (together with the wild buffalo) is the gaur (Bos gaurus). A good adult bull may weigh a ton and stand over 6 feet at the shoulder. Gaur is also more commonly known as the Indian bison, however this is not correct. Gaur has an extensive range covering the entire Central Indian plateau and extending in a north-easternly direction along the foot-hills of the Himalaya spreading to Burma and Malaya. In Malaya it is called the seladang, which is probably the same animal with another name. The range of gaur in India is now confined to the Central Indian plateau. Chota Nagpur area of Bihar and parts of Orissa. In the South, it extends into the Western Ghats. The most famous sanctuary for gaur is the Bandipur Madumalai complex in the states of Karnataka and Tamil Nadu. Gaur is also a protected species and is considered rare. However, it has been subject to periodic large scale mortality due to rinderpest or foot and mouth disease communicated through domestic cattle which graze in the same range. Inoculation of the sanctuaries containing gaur are the two steps being taken by the wildlife authorities to protect this animal.

The Indian Desert

In Rajasthan just east of the Indus Valley, lies the Thar or the Indian Desert. This desert is linked through Sind and Baluchistan in Pakistan to the deserts of Arabia and the Middle East and many of the animals of the Indian desert have historical affinities with the Ethisopian realms. This area was not always a desert, in fact archaeological finds in the region have revealed the existence of a flourishing civilization in this area some 5000 years ago, now known to Historians as the 'Indus Valley Civilization'. Man's poor agricultural practices, deforestation and overgrazing by his livestock probably destroyed the civilization and turned the region into an arid plain where few wild animals could survive. Even today the nomadic tribes in Rajasthan are mainly pastoral and maintain large herds of sheep and goats. The process of deforestation and desertification continues in the desert fringe and the process has accelerated within the last century so much so that the scientists have estimated that the desert is extending around its periphery in the north-west, at the rate of nearly half a mile per year or roughly 8 km every decade.

Water is the single biggest problem in the desert. The only sources of water are the oasis of desert springs, which are ground water reservoirs coming to the surface through artisan wells or the natural exposure of the water bearing strata. Large desert waste lands however stretch for mile after mile without any signs of water, Daytime temperatures rise very high reaching 48° C or more while night temperatures often fall to freezing point.

Both plants and animals face the problem of maintaining the water balance of their bodies under these highly adverse conditions. Sources of water are limited, while water is lost through body tissues in a number of ways. Plants lose water through high transpiration while the animals lose water in several ways; by evaporation from the surface of the skin, from the moist respiratory surface during breathing; through urine and faeces and in the mammals, with the milk they feed to their young. To maintain their water balance, water loss must be offset by water intake.

Plants have reduced transpirational losses through various modifications of plant parts. Desert trees are mostly thorny with highly reduced leaf surfaces. Roots go deep in search of water. Many plants have modified stems; leaf surfaces have become thick and oily. Cacti and succulents are the principal plant species in the desert.

Animals in the desert have become specialised in many ways. A large number of desert mammals are burrowing animals that live underground. The extreme temperature variations measured at the surface of the desert do not persist to any great depth and a burrow one metre underground enjoys a fairly constant temperature of about 30°C. Thus by living underground high temperature variations are easily avoided.

Desert vertebrates also have remarkable physiological adaptations to conserve water. They have efficient kidneys which secrete highly concentrated urine and specialised intestines that removes water from the faecal material before it is excreted.

Rodents probably represent the largest group in the mammalian fauna of the desert. The Indian desert gerbils are burrow dwellers. They are pale grey mouse like rodents with long tufted tails and live together in small colonies. Their faeces are dry and their urine very concentrated. Perhaps their most extraordinary adaptation is their ability to live on perfectly dry food with no water at all. All their body fluids is derived from metabolic water.

Perhaps the largest mammal of the Indian desert is the Asiatic wild ass. It lives in the southwest of the Thar desert on the salt flats of the Little Rann of Kutch. This area is partly inundated by about 1/2 m or so of water during the monsoon. However during the dry months from November to June there is no water and the islands of the wet season become small hilocks locally called 'bets' supporting some grass and little trees similar to the desert scrub of the mainland. Throughout this semi-desert the wild ass roams. These animals have a great tolerance to dehydration and can withstand a water loss of 30 per cent of their body weight. The ass restores its water condition to a normal level by drinking where water is available, even faster than a camel.

Another ungulate of this region is the blackbuck though strictly speaking it is an amimal which lives on the border of the desert in semi-arid conditions. Once blackbuck was the most abundant wild ungulate of India. It is an animal of the open terrain inhabiting the scrab forests and open savannah. Now it is adapted to the wastelands in cultivated areas and strangely enough in spite of damage to the agricultural crops by this animal it is protected by the villagers through religious beliefs. One such tribe, the Vishnois of Rajasthan afford full protection to blackbuck and would not even hesitate to kill anybody daring to hunt this animal.

Carnivores belonging to the cat and dog family also occur in the desert. Amongst these are the desert cat and the caracal belonging to the cat family, specially adapted for desert conditions. A sub-species of the red fox known as the desert fox also inhabits the desert regions and preys on small rodents and birds.

Reptiles are also fairly well represented in the desert. Being poikilothermous (cold-blooded) they can adjust to the desert heat by adjusting their exposure. to sun and shade according to temperature. Of the four living orders of reptiles, three-snakes, lizards and tortoises are well represented in the desert. A number of desertlizards (agamids, lacertids and a variety of geckos) use the heat of the sun's rays to warm their bodies. When warm, they avoid further exposure by seeking out shade, or burrowing under the sand. They are diural and can move swiftly over the hot ground to catch insects and avoid hawks. Many have toes fringed with small elongated scales which provide a snowshoe effect when the animal runs over loose sand. Snakes are more nocturnal and avoid the day-time heat. The saw-scaled viper hunts at night, seeking lizards and small mammals especially geribils whose burrows it invades. These reptiles from desert areas excrete most of their nitrogenous waste as uric acid in the form of thick white paste so that they lost little water in this way.

Amongst the birds of the desert the best known and the most widely discussed in the present times (including a recent international symposium organised exclusively for it) is the Great Indian bustard. It is a large beautiful bird which inhabits dry open country. It is a large beautiful bird which inhabits dry open country. It can fly quite well but prefers to run on the ground. The terrain which it inhabits is the desert fringe consisting of extensive stony grasslands. The bustard is omnivorous eating seeds as well as insects and even small vertebrates when it can catch them. Unfortunately these spectacular birds have been unable to withstand human pressure and have become seriously endangered. The habitats of the Great Indian bustard have been mostly destroyed and they have been widely hunted as a table bird, recently, the bustard was the subject of considerable public outcry due to bustard hunts with falcons by some Sheikhs from the Middle East.

The Tropical Rain Forest Region

The region consists of the heavy rainfall zones. These comprise the north-eastern India (Khasi jaintia hills and the lower Himalayan slopes embracing areas of Arunachal Pradesh, Assam, Meghalaya, Nagaland and Tripura) and the Western Ghats region in the South including the Malabar coast.

The Nilgiris, an off-shoot of the Western Ghats, rise precipitously to form extensive grassy downs and tablelands interspersed with densely forested gorges of evergreen vegetation known as sholas. Sholas, similar to those of the Nilgiris occur in the Annamalai and Palni hills also and other South Indian regions. They provide the main shelter to wild elephants, gaur another large animals of these hills. The high level forests of the Nilgiris also have affinities with the high altitude forests of the Assam ranges. Many of the trees found in these high sholas and some forms of animal life are common to both areas. Characteristically, Himalayan animal as the tahr (called the Nilgiri tahr), the pine-marten and the European otter occur in the Nilgiris.

Then rain forests of the Western Ghats and the eastern Himalayas consist of very dense and lofty trees with a multitude of species occuring in the same area. Hundreds of species of trees can be identified in a hectare of land, not to mention the mosses, ferms epiphites, orchids, liannas and vines, herbs shrubs and fungi that make up this region the most diverse habitat. Giant trees stretch up towards the sun. Buttress roots, anchored within the soil, support the smooth straight trunks which rise 30 mts. or more before branching out.

The spreading crowns effectively block out most of the light from the ground beneath. Dipterocarpus supp predominate in these forests and this type of vegetation is often called Diperocarp forests.

Other smaller plants gain access to the life giving day light by clinging to the branches of the giant trees. Many of these epiphites have roots and tubers adapted for collecting and storing moisture from their host when there is rain. This region is therefore, the home of a large variety of a tropical orchids, particularly the North Eastern region of Arunachal Pradesh and Sikkim.

Working vertically down from the canopy, the rain forests can be divided into 3 horizontal layers—the emergent trees in the upper storey constitute the top canopy which receives most of the available sunlight. Below this is the middle stroey of trees of slightly lower height composed of species which are tolerant of shade and the young saplings of the taller dominants. On the ground there is a carpet of fallen leaves and decaying wood with occasional tangles of canes forming impenetrable cane brankes.

Within this rich environment live a host of animals of all kinds. On the ground only big herbivores, i.e., elephants or solitary deer like the barking deer can survive. But most of the species in this region are tree dwellers. Over millions of years, animals have evolved in the tropical rain forests as specialised climbers and fliers spending all their lives in the canopy, and specialist feeders always eating fruit, and perhaps fruits only from a limited range of plant species. Amongst the tree dwellers, the most prominent are the group of non-human primates. The North-east region of Arunachal Pradesh and Assam is inhabited by the hoolock gibbon (is the only ape found in India at present); the golden langur (an animal discovered only in 1950s by the late E.P. Gee, in the forest of Manas Reserve), the capped langur or leaf monkey, the Assam macque and the pigtailed macaque. In the South, the most prominent primate of the evergreen forests is the lion-tailed macaquea species which is now considered rare and endangered and is specially protected. The other primates of this region are the Nilgiri langur and the slender loris, while the slow loris inhabits the forests of north-eastern India in Assam where it is seldom seen, being purely nocturnal and is reported to spend the day in sleep curled like a ball, preferably in some dry holes or crevice.

Bats are wide-spread group of flying mammals and a host of them dwell in these tropical rain forests. Other climbing animals and tree dwellers characteristic of this region are the giant squirrel and the civets. The binturong and the red panda (both rare) are characteristic of the evergreen forests in the Eastern region.

A number of climbing animals have evolved gliding mechanisms and are particularly characteristic of these forests. Among these are the flying squirrels. The other characteristic species of the Western Ghats are the Nilgiri mangoose, the stripnecked mangoose, the malabar civet and the spiny mouse.

The flora and fauna of these evergreen region have not been fully explored. Being a store house of a large variety of a plants and animals, these forests represent one of the richest gene pool resources of flora and fauna in the country. Though a large number of such forests have not been destroyed for various plantation crops like rubber, cocoa, coffee etc., whatever virgin forests remain have to be specially protected as in the Silent Valley of Kerala or the rich orchid belt of the Northeastern Himalayas in Arunachal Pradesh and Sikkim. The government of

India is actively considering declaring these areas Biosphere Reserves.

No description of India's Wildlife will be complete without mentioning those in three specially interesting habitats :

(1) The wildlife in the Andaman & Nicobar Islands.

(2) The water birds in the Bharatpur Sanctuary.

(3) The mangrove forest of the Sunderbans.

(1) The wildlife in the Andaman & Nicobar Islands

The Andaman & Nicobar Islands lie in the Bay of Bengal on a curve between Cape Negrais in Burma and the west end of Sumatra. These consist of 348 islands of various sizes which cover an area of 8,136 sq.km. The most distinctive features of the terrain are the hills lying north to south along the general direction of the islands. The highest point is Saddle Peak 800 mt. high, in the north Andamans. The climate of these islands is tropical and the islands receive both the southwest and northeast monsoons with an annual precipitations of 2750 to 4550 mm occurring for 9-10 mouths in the year, the only dry months being February and March. These islands are the home of tropical rain forests, and in the words of Sir H.G. Champion "Forests in their pristine glory, if found anywhere in Southeast Asia, it is in the Andamans". Giant *Dipterocarps, Terminalias* and the Andaman Padauk characterise these forests.

Bode Kloss (1903), a naturalist who studied the fauna of the islands at the turn of the century, speculated that the islands derived the bulk of their species from the distant Indian region with the Indo-Burman and Indo Malaysian regions being represented to a lesser degree, while Ferrar (1948) maintains that the fauna of the Andamans is more closely aligned to that of Burma than to that of India while the fauna of Nicobar has some affinities with that of Malaysia. Long periods of isolation have produced biotopes which have helped to evolve and sustain differences in the animal populations in the different islands and the group presents a wonderful field of study and research.

There are about 35 species of mammals, a number of interesting reptiles and a variety of marine fauna. The most noteworthy features of the mammalian fauna are the preponderance of bats (16 species) and rats (13 species) which together constitute nearly 3/4th of the total number of mammals known to occur in the islands and the absence of nearly all representative of ungulates, squirrels, carnivora and mammals, which are characteristic of the surrounding regions and abound on other islands of equal distance from the mainland of India and Burma.

The Andaman pig is one of the large mammals found in the Andamans. This is smaller than the one found on the mainland. The crab-eating coamaque, the palm civet and various species of deer are also important larger mammals. Most of these deer species were introduced into the Andamans during the 1920s and included spotted deer, barking deer, hog deer and sambar. All of these, with the exception of Sambar have survived to varying degrees and the spotted deer in particular has thrived beyond expectations. The plan civet, also introduced in these islands, has survived in most places and has become a serious pest to poultry and crops such as banana, cocoa, coffee etc.

Among the important marine mammal found around the islands are the dugong (after which the Dugong Creek in Little Andaman is named), the false killer whale, and the dolphin. The dugong is a rare and threatened animal and is completely protected.

The avifauna of these islands, which has recently

been studied by Humayun Abdulali, is reported to have a total number of 240 species and sub-species of birds. There are roughly 92 migrants, 32 residents, and 112 resident endemic species. Among the endemic species are such rare species like the Narcondum hornbill, the Nicobar pigeon and the megapode which has Australian affinities and incubates its eggs by placing them in an earthern mound.

The other interesting birds found in the area include white bellied sea-eagle, the whitebreasted swiftlet, and a number of fruit pigeons.

In the tidal creeks and mangrove swamps all along the coastal areas of the Andaman and Nicobar, resides the Salt-water crocodile which is again a threatened and completely protected species. A number of marine turtiles are indigenous to these islands. The green turtle is quite common and breeds in profusion on South Sentinel Islands. The robber or coconut crab has so far been recorded only on the south Sentinel Island. It is the only orthropod included in the Schedule 1 of the Wildlife (Protection) Act as a rate and endangered species. Recent studies on this crab by Altervoet and Davis (1976) have greatly contributed to our knowledge of this interesting animal and have indicated possibilities of farming these crabs for commercial purposes.

A number of Lizards occur in these areas, of which the largest is the water monitor. The green lizard is also peculiar to these islands. There are about 40 species of snakes, among which the poisonous species include the cobra, viper, coral and sea snakes, Pythons can be found in the Nicobar group.

(2) Water Birds in the Bharatpur Sanctuary

Situated at a distance of 156 km from Delhi and Jaipur and only 50 km from Taj Mahal at Agra, near the town of

Bharatpur is a virtual craven for water birds called the Keoladeo Ghana. The fresh water swamp of Keoladeo Ghana with the picturesque Kadam trees, stands of Acacia and profuse aquatic vegetation extending over an area of 2,900 hectares is the habitat of rich micro and macro living forms including millions of snails, fish, frogs and crustacea. This abundant food, renewed each year by flood waters of the Gambhiri and the Banganga has been attracting thousands of resident water birds to nest and breed every year. Over a million migratory ducks teals, geese including waders have also been attracting thousands of resident water birds to nest and breed every year. Over a million migratory ducks, teals, geese including waders have also been coming here during winter from far distant temperate regions of Eurasia, making it a truly international community of birds, unique in its richness and variety.

This spectacular area had been the famous shooting preserve of the Maharaja of Bharatpur, where grandiose duck-shoots were regularly organised for British Viceroys and other dignitaries, establishing many a shooting record. Thousands of ducks were shot in a single shoot. Now this unique bird habitat has been converted into the 'Bharatpur Sanctuary' and is an area of great interest to ornithologists and photonaturalists as well as tourists from all over the world.

In a way the whole sanctuary is one lake. It is however divided into a number of smaller lakes separated by boundary roads. Water level in the lake is maintained to a depth of 1.5 mt. through gates. The water spread during the monsoon and winter is about 500 hectares. Trees of Acacia nilotica, half submerged in water or planted over mounds with myriad nesting birds on tree tops on either side of a bund provide a unique spectacle. Nearly 328 species of birds have been identified in this region, of which 110 are migratory birds. Amongst the breeding birds in the Bharatpur Sanctuary, the more interesting are the painted stork, large egret, medium or smaller egret, little egret, openabilled stork, spoonbill, the white ibis, the pond heron, the night heron, the snake bird or Indian darter, the large cormorant, and the small cormorant. Most of them can be seen in mixed heronries often on the same tree. The iarge cormorant, and the small cormorant. Most of them can be seen in mixed heronries often on the same tree. The large and dainty painted stork occupies the top crown of the babul trees, while the other birds, spoonhill, ibises, herons, occupy lower positions in a systematic gradation. During the height of the breeding season, it is a feast of bird music, noise and motion everywhere. All the time there are birds taking flight or returning to their nests with nestlings twittering, chirping and demanding food. Cormorants form hunting parties in the water, driving fish, wading and catching them. Amongst the winter visitors, there are a number of geese and ducks, rosy or white pelican. Dalmatian pelica, bar-headed geese, grey lag geese, pochards, pintails, sandpipers, a large numbers of warblers, wagtails and pipits. However, the VIP visitor to this area is the Siberian crane which is one of the most rare birds. This picturesque bird stands upright on long legs to an average height of 4.5 ft. Its plumage is white except for glossy black wing tips. Its face has no feathers, red skin surrounds startling yellow eyes. Today it is estimated that no more than 400 Siberian cranes survive world wide. These consist of two breeding populations. Most breed on the Yakutian tundra in Eastern Asia and migrate to South to some unknown wintering grounds in China, perhaps along the Yangtse river. There is a smaller flock which breeds somewhere in the western parts of Russia, in some impenetrable regions of the Obe river. It is this flock which travels 5,000 miles via a lake in Afghanistan, Ab-i-Estada, to the Bharatpur Sanctuary in India to spend the winter. In 1979, the number of this crane, was estimated at only 41. A water sedge Cypress rotundus found in the swamps of Bharatpur is the main attraction for the Siberian crane, which is a vegetarian; the tuberous roots of this sedge forming its principal food.

Member of the International Crane Foundation are busy trying to save this rare bird and though they are perfectly safe in Bharatpur, there are dangers along their travel route in Afhansistan and Iran over which there is a very little control.

Bharatpur has also been a centre of studies of migratory patterns of birds conducted by India's most famous ornithologist, Late Dr. Salim Ali, under the aegis of the Bombay Natural History Society. A number of interesting studies on migratory patterns of birds have been undertaken in this sanctuary through bird banding operations. A project on the study of wetland ecology is also under consideration of the Government of India in this sanctuary under a collaboration between the U.S. Fish and Wildlife Service and Bombay Natural History Society.

(3) The mangrove forest of the Sunderbans

Mangrove swamps characterise the estuaries of many of the larger Indian rivers which drain to the Bay of Bengal. However, in India, the most well-known mangrove forests are found in the Sunderbans delta of the river Ganges where both the Brahmaputra and Ganges join and drain into the Bay of Bengal. Here the river has divided itself into hundreds of streams to form narrow creeks. In between these creeks, lie the mud flats of the delta region which are periodically flooded through tidal inundations by the sea. Great quantities of silt are brought down by the river each year and deposited into the sea to claim more land from the shallow sea. Watr logging, humidity, salinity and regular inundations are the salient features of this habitat. From it seaward to its landward limit, mangrove swamps can be divided into 3 distinct zones-(1) the low or the freshly formed land which remains mostly below the tide level, (2) the slightly higher land which is frequently inundated by salt water, and (3) the higher ground generally above the tidal level (these areas go under water once or twice during May to October on full and new moon or at exceptionally high tides.)

The lower tidal zones are characterised by the primary colonisers, the *Sonneratia* and *Avicennia* trees. These trees have extensive underground root which bind the mud together. The roots show great tolerance to high salinity as they are mostly under water. Since no air can penetrate this aline mud, the *Sonneration* roots throw up vertical respiratory schoots (Pheumatophores). Above this zone is the zone characterised by *Rhizophora, Bruguiria* and *Excaecaria Careops* forest. This forest occupies slightly higher ground and is only partially inundated. The Rhizophora are the familiar tangle rooted mangroves. Each tree is supported by a mass of branching. Stilt roots, which like the *Sonneratia* roots can respire when exposed at low tide. The *Exacecaria-Cereops* forest is the most common and the principal association in the Sunderbans delta covering nearly 70% of the forest. Above this level is the high land rarely inundated, supporting forests of Phoenix, either pure or in association with *Excaecaria* and in the highest portion we have the *Heriteria* forest with thick under growth of *Phoenix* and Neepa palms. In some places *Excaecaria* forms a high portion of mixture.

The animals community in these mangrove forests makes the environment exceedingly complex. At low tide, the exposed mud flats in the lower portions are invaded by an army of small crabs and fish. These fishes have special adaptations which make them independent of the water. They are the mud-skippers or semiterrestrial gobies; they can leap with a flick of the tail or even swim across the mud or water by a series of hopes but normally they move slowly crawling or 'crutching' on their pectoral fins. Later they revert to the normal fish mode of swimming and respiration. The most spectacular feat of the mud-skipper, however, is its ability to climb up the roots of mangrove treees which presents the strange sight of tree-climbing fishes.

Crowds of small active crabs settle over the mud flats as the tide recedes. The most famous are the land crabs, and the fiddler crabs whose males have huge brightly coloured claws. All these crabs live in jealously guarded burrows. As soon as the tide ends they emerge, rolling out balls of mud and sand between their burrows. These are their escape holes into which they scamper, if danger threatens in the form of storks, herons, monkeys or domestic ducks. With high tide, they retreat into their holes and block out the entrance to keep out the water trapping an air bubble which enables *them to breathe. One small crab,* the Dorippe has two pairs of legs modified as little claws that carry an oval plate on which a small sea aneomone lives. The arrangement is mutually beneficial. The anemone is a firm base for attachment and also benefits from the scraps of food dropped by the crab. In return, the crab gains protection from the anemone as it has nasty sting which discourages the predators from attacking this combination.

Above the mud flats on the mangrove trees themselves live numberous weaver ants which make their nests in trees by weaving together 5,6 or more leaves with their fine silken thread. This nest guarded by black solider ants, houses the colony's strange green queen. Columns of red workers scavenge over ground and bushes for food.

In the higher regions of the mangrove, the fauna changes and typical land animal like the spotted deer, and the pigs, monitor lizards and monkey characterise the area. Spotted deer are found the abundance and form large herds. Earlier, the jawan rhinoceros, swamp deer and wild buffaloes were also a feature of the Sunderbans. They however no longer exist in the area.

However, the most interesting animal of the Sunderbans is the Sunderbans tiger which has adapted so intriguingly to this strange habitat of high salinity and periodic inundation, leading an amphibious life, subsisting on a variety of food which includes, apart from the normal prey of spotted deer and wild boar, even fish and crabs. The most peculiar trait of the Sunderbans tiger is its tendency to man-eating. It is believed that all sunderbans tiger are man-eaters and every year, a large number of fishermen, honey-collectors and other villagers fall prey to these tigers. The tigers are known to swim in the creeks and climb into boats to take their prey or lurk in thick bushes of the *Phoenix* palm to ambush the honey-collectors and other villagers fall prey to these tigers. Whether this man-eating tendency is acquired because of the peculiar habitat and the vulnerability of the fishermen moving in their narrow boats in creeks for fish or honey-collectors moving in the mangrove forests with poor mobility or whether there is any ecological correlation between its saline habitat and the food habits of the tiger, has not yet been determined. However, it is doubtless a fact that there is a heavy annual toll of human beings in the area.

2 Endangered Species of India

The increase of human population in the last few decades demanding development in various spheres has resulted directly or indirectly in the sudden and often far-reaching disturbances in natural ecosystems. The growth of large urban areas, construction activities such as dams, buildings and roads, encroachment on vast areas of forest lands for extension of arable expanses and mining operations are examples of direct onslaughts on vast areas of forest lands for extension of arable expanses and mining operations are examples of direct onslaughts on nature which have steadily depleted natural resources. Besides, the pollution resulting from urbanization, industrialization, excessive use of chemical fertilizers and p sticides and also of the several giant aircrafts have in no small way contributed to eco-imbalances in soil, water and air. These imbalances have posed problems for the very survival of wild life. Besides these, in order to satisfy his ego and established supremacy over other life by collecting hunting, trapping and poaching man has driven certain species on the verge of extinction.

Extinction has been the destiny of a great number of plant species including several unique and irreplaceable varieties. Some of these have disappeared from the earth in nature's own process of evolutionary change. But for many other, extinction has been caused by man because of his inadvertence as well as ignorance about their economic potential and ecological functions. So far only a small fraction of the diverse population of plants is known to us and a large majority still remains unidentified. Even in respect of familiar species in many cases, little more is known than their appearance and location.

Through indiscriminate exploitation, destruction of habitats, spread of harmful chemicals and intorduction of alien species, a number of plants have already disappeared while others await a similar fate, at times, even without our being aware of their existence. Thus human interference has been, more often than not, responsible for the depletion of plant resources and consequent decline of genetic diversity. According to an estimate of the Threatened Plants Committee of the International Union for Conservation of Nature and Natural Resources (IUCN) about 10% (20,000 to 30,000) of the world's flowering plants are reported to be "dangerously rare" or "under threat". Norman Myers, noted conservationist, concludes that at least one species is disappearing each day in the tropical forests alone and in a few more years there may well be a species lost each hour. The seriousness of the situation can be understood from the observation of Peter H Raven that a "disappearing plant can take with it 10 to 30 dependent species, such as insects, higher animals and even other plants."

Realizing the ecological importance and economic utility of plant resources, there is growing concern throughout the world about the need for conservation. As pointed out by Otto Frankel, the well known Australian geneticist, it is now time for determined efforts "to keep evolutionary options open so far as we can and to resist from eroding the richness of genetic reserves. During the recent past such concern has found expression in the establishment of an International Board for Plant Genetic Resources, and the UNESCO launched the Man and the Biosphere (MAB) programme.

Threat to plant species

The Forest administration in India is more than 100 years old. Efforts to identify plant species as a part of wildlife and to recognise the importance for providing special protection to endangered plant species have so far been negligible. While the rapid disappearance of some beautiful animals stimulated efforts to protect fauna, this has not been the case with flora. The word 'Wildlife' had till recently been considered synonymous with animal life and consequently, conservations and naturalists have their attention only to conservation of animal species.

It was only in the year 1968 at an International Conference (UNESCO 1968) that the problem of conservation of flora & fauna was appreciated and several recommendations were made urging the International Biological Programme (IBP), the International Union for Conservation of Nature and Natural resources (IUCN) and various international and national organisations to initiate studies in to the problems involved, particularly the problem of protecting and preserving wild fauna and flora in their natural habitat/ecosystems establishing nature reserves.

Later at the 10th General Meeting of the IUCN, the Survival Service Commission reviewed the status of endangered species of plants and their habitats. The recent promulgations of the United States Endangered Species Act (1973), the UK Wild Creatures and Wild Plants Act (1975), development of international conventions on conseervations (Wetlands Convention) and the setting up of Biological Records Centre of the Nature Conservancy UK, and threatened plants, orchids, cycads treewide awakening of the need for the conservations, preservation and protection of floras of the world.

According to the Volume 5 of Red Data Book on angiosperms started by the Survival Service Commission of the IUCN in the year 1970 (Melville, 1970), it is estimated that out of the total of three lakhs (0.3 million) species of plants in the world, over 20,000 were in the category of either endangered or vulnerable and threatened with extinction by the year 2000 A.D.

Rare species of plants, their categories and definitions

The following categories of rare plants have been recognised by the IUCN, mainly based on (i) present distribution, (ii) decline in numer of time, (iii) abundance and quality of natural habitats, and (iv) biology and potential value of the species. These are defined as follows for purposes of conservation:

Endangered (E)

Taxa in danger of extinction and whose survival is unlikely if the causal factors continue operating.

Included are taxa whose numbers have been reduced to a critical level or whose habitats have been so drastically reduced that they are deemed to be in immediate danger of extinction.

Vulnerable (V)

Taxa believed likely to move into the endangered category in the near future if the causal factors continue operating.

Included are taxa, of which most or all the populations are decreasing because of over-exploitation, extensive destruction of habitat or other environmental, disturbance; taxa with populations that have been seriously depleted and whose ultimate security is not yet assured; and taxa with populations that are still abundant but are under threat from serious adverse factors throughout their range.

Rare (R)
Taxa with small world populations that are not at present endangered or vulnerable, but are at risk.

These taxes are usually localised within restricted geographical areas or habitats or thinly scattered over a more extensive range.

Threatened (T)
Threatened is used in the conservation context for species which are in one of the categories Endangered, Vulnerable and Rare. Some species are mrked as threatened where it is known that they are Endangered, Vulnearable or Rare, but there is not enough information to say which of these three categories is appropriate.

Out of danger (O)
Taxa formerly included in one of the above categories, but which are now considered relatively secure because effective conservation measures have been taken or the previous threat to their survival has been removed.

Indeterminate (I)
Taxa that are suspected of belonging to one of the first three categories, but for which insufficient information is currently available.

Main causes of threat to species
It is well known now that several plant species have become extinct due to certain natural phenomena, such as land upheavals, volcanic eruptions, glaciation, protracted periods of rain or drought, spreading of desert lands, forest fires, and eutrophication in the geological past. While such natural processes in the past had no

doubt led to the extinction of flora, the. resulting new environmental conditions had also resulted in the evolution and speciation of new flora and migration of floral elements.

But, in recent times man with his anthrpopogenic associates and other factors or practices such as fire or 'slash and burn' for shifting cultivation (also called as 'jhum' or 'podu' cultivation in India), grazing by cattle and by several other mechanical means, has suddenly accelerated disastrous condition in natural ecosystems.

Besides, commercial exploitation of entire plants, roots, rhizomes, tubers bulbs seeds and fruits has been the prime cause of depletion of more important wild economic plants throughout the world for lucrative financial gains, in the trade which flourishes both by legal and illegal means. *Rauvolfia serpentina, Coptis teeta, Dioscorea spp,* and *Podophylum; hexandrum* serve as good examples.

Area vulnerable to rarity and extinction of species in India

In India, the maximum threat to plant species lies in areas which are sujected to considerable development activity e.g. forests which lie in close proximity to villages and agricultural lands. Such areas occur in the entire length and breath of the Himalayas, the Western and Eastern Ghats of peninsular India, the Vindhya and Satpura Hill ranges, and the densely forested hilly region in Assam, Nagaland, Meghalaya, Minipur and Mizoram in Eastern India. These regions incidentally abound in many tourist spots and pilgrimate centres, which attract large number of people who cause damange to the flora.

Besides, endemics which are confined to narrow, restricted areas or specialized ecological niches, also in some respects fall in the category of threatened plants. Some phytogeographers believe that endemic species or genera are the survivors of the one time wisely distributed groups, which are now in the course of gradual extinction. Others maintain that they are newly evolved and recent forms still in the process of extending their geographical range of distribution. In such cases, due to their rarity they are to be considered as threatened plants, because any disturbance or imbalance in their narrowly confined ecosystems/habitats could mean extermination of the species.

Isolation, as in the case of island floras, and natural barriers such as mountain ranges are considered main factors resulting in endemism.

India has a very rich element of endemics in its flora. According to Chatterjee (1940) Indian subcontinent has about 61.5 per cent to endemic flora with about 7,000 endemic species and 134 endemic genera. Of these the Himalayas and the Khasi Hills account for about 3,000 and Deccan peninsula for about 2,000 endemic species. These figures speak eloquently of the great need for protecting the endemics. This can be done effectively by carefully analyzing the floristic composition of the various phytogeographical units of India flora and by selecting suitable natural forests in these units and preserving them as 'Biosphere Reserves'.

Sapria himalayana, Uvaria Iurida, Alcimandra cathcartii, Magnolia gustavii, M. pealiana, Pachylarnax pleiocarpa, Nepenthes khasiana, Dicentra royler, several species of *Primula* and *Rhododendron* and the Lady's Slipper orchids. *Paphiopodilum insigne, P. hirsutissimum, P faireenum, P spicerianum* and *P. venusturm,* are some of the noteworthy endemics of the Himalays and Khasi Hills. *Antiaris toxicaria, Campanual cytinoides, Pedicularis perroter* and the some species of the Podostemaceae are endemic to Western ghats or the Nilgiri Hills in South India.

Himalaya and Eastern India

1. *Abies delavay, Franchet (Pinaceae)* Tree A Chinense Fir recorded from Arunachal Pradesh, rare occurrence.
2. *Acanthephippum sylthetense* Lindi (Orchidaceae) Herb A terrestrial orchid distributed in Garampani sanctuary (Assam Khasi, Orissa, Abor Hills and Sylhet (Bangladesh) rare occurrence.
3. *Aconitum deinorrhizum* Staph (Ranunculaceae) Herb One of the important Indian Aconites of medicinal value originally discovered from Bashahr Himalaya and recently reported from Bhadrawah, J&K State. rare occurrence.
4. *Adinandra griffthu* Dyer (Theaceae) Tree Distributed in Khasi Hills, Meghalaya rare occurrence.
5. *Aglaia perviridis* Hiern (Meliceae) Tree Distributed in Khasi Hills, Meghalaya, rare occurrence.
6. *Amblyanthus multiflorus* Mez (Myrsinaceae) Shrurb Rarely reported from Assam.
7. *Amacppsa ilicoides* Mast (Olacaceae) Tree Distributed in Khasi Hills, Meghalaya, rare occurrence.
8. *Anoectochilus sikkimensis* King & Plant (orchidaceae) Terrestrial herb, the 'Jewel Orchid'. Distributed in Sikkim Himalaya, recently reported from Garampani sanctuary, Assam, rare occurrence.
9. *Angopteris evecta Hoffm* (Angiopteridaceae) Terrestrial fern. Widely distributed in India, Sri Lanka and sepals Asia, sporadic in occurrence.
10. *Aphyllorchis montana Reichb* (Orchidaceae) Herb A Terrestrial orchid, petals straw-coloured purple at back and tips. Distributed in Sikkim and Khasi Hills, rare occurrence.
11. *Arachnanthe cathcarti* (Lindi) Benth (Esmeralda cathcartii (Lindi) Reichb) (Orchidaceae)

Epiphytic Attractive with large brownish-red flowers. Distributed in East Nepal, Sikkim, Bhutan and Arunachal Pradesh, rare occurrence.

12. *Arachnanthe clarker* (Reichb ff) Rolfe (*Esmeralda clarker* Reichb f) (Orchidaceae) Epiphytic Robus orchid with large, attractive, yellowish-brown flowers. Occurs only in Sikkim and Bhutan, rare occurrence.

13. *Artemisia amygdalina* Decne (Asteraceae) Herb. A botanically interesting species with leaves like those of a willow (J.D. Hooker). Known only from Pir Panjal, J&K State, extremely rare in occurrence.

14. *Arundina graminifolia* (D. Don) Hochr (Orchidaceae) Terrestrial, perennial plant. The Bamboo orchid with highly ornamental flowers. Distributed in Sikkim, Khasi Hills and Arunachal Pradesh, rare occurrrence, horticultural value, exploited in trade.

15. *Astragalus strobiliferus* Royle (Papilionacea)—Herb. The only member of the Tragacanth group of Himalayan Asttragali occurring in W. Himalaya; rare occurrence.

16. *Atropa acumina* Royle ex Lindl. (Solanaceae)—Herb. The 'Indian Belladonna', distributed in Himachal Pradesh and Kashmir, rare occurrence; exploited as medicinal plant.

17. *Balanophora dioica* R. Br. (Balanophoraceae)—Herb. Clump-forming, fleshy, clubshaped parasite on treee roots in dense forests in the entire Himalayan region; of extreme rare occurrence; botanical interest (Also other species of Balanophora).

18. *Botrychium virginianum* Sw (Ophioglossaceae)—Perennial. A pretty fern of botanical interest. Distributed in the Himalayas (Kumaon to Aruncahal Pradesh), Khasi Hills, Peninsular India and Sri Lanka; rare occurrence.

19. *Brainae insignis* (Hook) J. Sm. (Blechnaceae)—Terrestrial fern. Distributed only in Khasi Hills in India; very rare; also occurs in Malay Peninsula.
20. *Camellia caduca* C.B.Cl.ex. Brandis (Theaceae)—Shurb. Distributed in Skkim, Bhutan, Arunachal Pradesh and Khasi Hills; rare occurrence; botanical interest.
21. *Catamixis baccharoides* T. Thoms. (Asteraceae)—Herb. An endemic species restricted to rock cliffs in U.P. Siwaliks; rare occurrence; botanical interest.
22. *Colchicum luteum, Baker* (Liliaceae)—Herb. Endemic to N.W. Himalaya; rare occurrence; exploited in trade for its valuable alkaloid Colchicine.
23. *Captis teeta wall* (Ranunculaceae)—Herb. Distributed in Mishmi Hillls, Arunachal Pradesh, rare occurrence; heavily exploited in trade for its medicinal rhizomes—'Mishmi Teeta' now under cultivation.
24. *Cyathea gigantea* (Wall ex Hook) Holtt. (Cyatheaceae)—The giant tree fern of botanical interest. Widely distributed in hilly regions of eastern and peninsular India, but nowhere abundant; exploited for starch by the local people; trunk used for growing epiphytic orchids.
25. *Cymbidium macrorhizon Lidl.* (Orchidaceae)—The only terrestrial leafless Indian Cymbidum. Distributed in N.W. Himalaya, Sikkim, Khasi and Naga Hills; rare occurrence; horticultural importance.
26. *Cypripedium cordigerum* D.Don (Orchidaceae)—Terrestrial herb, highly ornamental "lady's Slipper Orchid", once reported abundant at Gulmarg (Coverntry's Wild flowers of Kashmir, 1923), but has been sighted only once in the course of five visits (K.A. Sahni). Also scarce in Western

Himalaya and Tehri-Garhwal region; extremely rare; threatened due to trade.

27. *Cypripedium elegans* Reichb. f. (Orchidaceae)—Terrestrilal herb; flowers brownish, attractive. One of the ornamental "Lady's Slipper Orchids". Distributed in lachen Valley, Sikkim and eastern Tibet rare occurrence, threatened due to trade.
28. *Cypripedium himalacicium* Rolfe (Orchidaceae)—Terrestrial herb; flowers showy, reddish or purplishred. Distributed in Western Himalayan, Sikkim and Bhutan, rare occurrence; threatened due to trade.
29. *Cypripedium macranthon* Sw, (Orchidaceae)—Terrestrial hrb with attractive large brownish-purple flowers, Distributed from Garhwal to Sikkim; rare occurrence; threatened due to trade.
30. *Dendrobium densiflrum* Wall ex Lindl (Orchidaceae)—Epiphyte. A highly ornamental orchid distributed in the tropical and subtropical forests of Nepal, Sikkim, Arunachal Pradesh and Khasi Hills but nowhere common; exploited in trade.
31. *Dianthus coschemiricum* Edgew, (Caryophyllaceae) — Herb apparently endemic in Kashmir, rare occurrence. (D. Chinesis L, is cultivated in India for its perfume-yielding flowers).
32. *Didiciea cunninghamii King et Prian* (Orchidaceae)—Terrestrial herb. An interesting monotypic orchid originally found in Sikkim and recently collected in Garhwal Himalayas after a lapse of seven decades; rare occurrence; botanical interest.
33. *Dioscorea deltoidea wall*, ex kunth (Discoreaceae). Hioscorcuceae—Herbaceous climber, Western Himalaya; becoming scarce; threatened due to over-exploitation from the wild for its medicinally important tubers.

34. *Dipteris Wallichii* (R. Br.) Moore (Dipteridaceae) m—Perennial herb. A fern of botanical interest. Distributed in Khasi and Jaintia Hills; recently discovered in Subansiri, Arunachal Pradesh; rare occurrence.
35. *Dischidia benghalensis* Cole br (Asclepiadaceae)—An interesting climber with fleshy leaves. Eastern Nepal, Sikkim, Assam rare occurrence; botanical interest.
36. *Dischidia rafflesiana* Wall (Asclepiadaceae)—A Climber with pitcher-like leaves, Assam; rare occurrence; a botanical interest.
37. *Drosera burmanni* vahl (Droseraceae)—Herb. The wellknown insectivorous 'Sundew' plant. Also the other species D *peltata* Sm. Both distributed from Himalayas to the east and south, scarcely distributed; botanical interest.
38. *Drosera indica* inn. (Droseraceae)—Herb. In marshy habitats. Distributed in Deccan peninsula, upto Chota Nagpur, recently reported from Birbhum (W. Bengla); sporadic, botanical interest.
39. *Elaeocorpus prufifolius* (C. Muell.) *Mast* (Elaeocarpaceae)—Tree distributd in Khasi Hills and Manipur, rare occurrence; botanical interest.
40. *Eremostachys superba* Royle ex. Benth (Lamiaceae)—An attractive large herb on the slopes of Dehra Dun Siwaliks. Threatened due to crosion of habitat.
41. *Eria crassicaulis* Hook f. (Orchidaceae)—Epiphyte. An attractive orchid with long recemes and pale purple-banded flowers. Distributed only in Kashi Hills; fare; horticultural value.
42. *Galeola lindleyana Reichb.* f. (Orchidaceae)—Tall, attractive saprophytic orchid with tomentose yellow flower, Sikkim, Kashi and Naga hills. Also another species *G. faiconeri* HK. f. with deep yellow

flowers, is found only in parts of Arunachal Pradesh; rare occurrence; botanical interest.

43. *Gentiana kurroo* Royle (Gentianaceae)—A herb with beautiful, large purple-blue flowers. Western Himalaya, rare occurrence, medicinally important; exploited in trade.

44. *Hedysarum cachemirianaum benth* ex Baker (Papilionaceae)—Herb. An attractive plant Endemic in Kashmir, rare occurrence; botanical interest.

45. *Helminthostachys zeylanica* (Linn) Hook (Helminthostachyaceae)—Terrestrial, perennial fern. Distributed in submontane parts of north-east and peninsular India and deltaic Bengal, recently collected in plibhit district, U.P. sporadic in distribution, botanical interest.

46. *Helwingia hinalaica Hook.* f & Thoms ex C.B.C.I.: (helwingiraceae)—Shrub. An attractive plant with flowers. Eastern himalaya and Khasi Hills; rare occurrenc; of great botanical interest.

47. *Liex embelioides* Hook. f. (Aquifoliaceae)—Small tree. Distributed in Khasi Hills; rare.

48. *Lodes hookeriana* Baill (Lcacinaceae)—Woody climber with attractive penicles of yellow-red drupe's. Khasi Hills (and Chittagon Hills, Bangladesh); rare occurrence botanical interest; horticultural value.

49. *Lavatera kashmiriana* Camb (Malvaceae)—Herb. 'The Kashmir Hollyhock' with pink flowers; endemic in Kashmir, scarce.

50. *Lespedez a elegans* Camb (Malvaceae)—Endemic in Kashmir; sporadic in occurrence.

51. *Loropetalum Chinense* Olive (Hamamelidaceae)–Small tree, distributed in Assam and Arunachal Pradesh; Scarce.

52. *Magnolia griffithil* Hook.f. & Thoms (Magnoliaceae)—Tree Endemic in Upper Assam; rare occurrence.

53. *Magnolia gustavi King* (Magnoliaceae)--Large tgree. Endemic in upper Assam; rare occurrence.

54. *Magnolia pterocarpa* Roxb (Magnoliaceae)—Tall tree. Scarcely found in Assam and Arunachal Pradesh, Hutchinson (Evolution and Phylogeny of flowering plants, 1969) considers this plant to be peraps the most ancient species of living flowering plants. Of considerable botanical interest.

55. *Meconopsis betonicifiolia* Franchet (Papaveraceae) — Perennial herb. Blue poppy of the Eastern Himalaya, also found in S.E., Tibet, Northern Burma and Yunnan; very rate in the wild; exploited for trade; introduced into Cultivation in Key Botanic Gardens in 1924 and now growth in Europe.

56. *Nardostachys grandiflora* DC (Valerianaceae)—Perennial herb. The Himalayan spikenard with aromatic rhizomes (Jatamansi), Alpine Himalayas, rare occurrence; exploited in trade

57. *Nepenthes khasiana* Hook. f. (Nepenthaceae)—Climbing under-shrub. The Indian insectivorous pitcher plant. Endemic in Khasi and Garo Hills; rare; botanical curio. The populations have dwindled due to excessive collections by traders and students of botany.

58. *Lax nana Wall* (Oraceaceae)—Tree. Distributed in Nepal, Darjeeling, recently reported from Assam; rare and sporadic in occurrence.

59. *Ormisia glauca* Wall (Papilionaceae)—Tree. Distributed in Nepal, Darjeeling, recently reported from Assam; rare and sporadic in occurrence.

60. *Osmunda regalis* Linn. (Osmundaceae)—Perennial shrub. The Royal Fern. Scattered in Kumaon Eastern Himalayas, Khasi Hills, Pachmarhi Hills, and peninsular India, but nowhere abundant; botanical interest.

61. *Paphiopedilum druryi* (Bedd.) Pfitz. (Orchidaceae)—Terrestrial herb, flowers yellow-green with reddish-purple blotches and streaks. Endemic in Travancore Hills, W. Ghats; rare occurrence.

62. *Paphiopedilum fairleanem* (Lindl Pfitz. (Orchidaceae)—Terrestrial herb. The "Asian Lady's Slipper Orchid", with large attractive greenish or creamy, reddish and purpple-veined flowers. Endemic in Kameng (Arunachal Pradesh) and Bhutan; very rare occurrence threatened due to habitat destriuction and indiscriminate collections for trade.

63. *Paphiopedilum hirsutissum* (Lindl) Pfitz. (Orchidaceae)—Terrestrial herb with large violet-purpel flowers of ornamental value. Endemic in Khasi Hills, becoming extremely rare due to top indiscriminate collections from the wild.

64. *Paphiopedilum insigne* (Wall) Pfitz. (Orchidaceae)—Terrestrial herb with attractive greenish, purple spotted flowers. Sparsely distributed in E. Himalayas, Khasi Hills and Nepal; becoming very scarce due to habitat destruction and exploitation in trade.

65. *Paphiopedilum spicerianum* (Reichb. F.) Pfitz. (Orchidaceae)—Terrestrial Scapigerous herb; flowers with purplish sepals and yellow or olive green, reds-spotted petals and brownish-crimison tinged lip. Endemic in cachar Hills, Assam; rare occurrence.

66. *Paphiopedilum venustum* (Lindl) Pfitz.

(Orchidaceae)—Terrestrial herb with blotched leaves; flowers howny and ornamental with white dark-striped sepals and greenish-purple wartypetals distributed in tropical Sikkim Himalayas and Meghalaya, becoming rare due to clearing of forests and over-exploitation from the wild.

67. *Paphiopedilum villosum* (Lindl)Pfitz. (Orchidaceae)—Terrestrial herb, flowers large, ornamental purplish-brown, distributed in Mizoram (Lushal Hills) and Moulmein, Tenasserim (Burma); extremely rare and threatened due to indiscriminate collections from the wild.

68. *Phyllostachys bambusoides* Sieb & Zucc (Poaceae)—Shrub. A Sino-Japanese bamboo distributed in Arunachal Pradesh; rare occurrence.

69. *Picea brachytyla* Pritz (Pinaceae)—Tree. A Chinense spruce with restricted distribution in Arunachal Pradesh; also occurs in N. Burma, Western Hupeh and W. Yunnan (China).

70. *Platycerium wallichill* Hook (Polyphyllaceae)—Epiphytic fern. Recently collected from Manipur, very rare; earlier known from Malay Peninsula, Tenasserim.

71. *Pleione humilis* D. Don (Orchidaceae)—Herb A pretty orchid distributed in Khasi Hills and Arunachal Pradesh, recently reported from Kumaon, W. Himalayas; rare occurrences; exploited in trade.

72. *Podophyllum hexamdrum* Royle (Podophyllacea)—Herb. The Himalayan many-appel widely distributed, but nowhere abundant. Heavily exploited for its medicinal value.

73. *Populus gamblei* Dodo (Salicxceae)—Tree. 'Poplar distributed in Eastern himalayas, rare.

74. *Potameia paradcxa* (Hook) Koster (Lauraceae)—Tree. A curious plant with 1-celled anthers; only once collected from Bhutan; rare occurrence; botanical interest.

75. *Psilotum nudum* (Linn). Beauv (Psilocaceae)—Perennial herb. A fern-ally of botanical interest. Widely distributed in India but nowhere common, botanical interest.

76. *Rauvolfia serpentina* (Linn), Benth ex Kruz (Apocynaceae)—Shrub. A very important medicinal plant of widespread distribution. Now, scarece in the wild due to over-exploitation. cultivated.

77. *Renanthera imschootiana* Rolfe (Orchidaceae)—Epiphytic herb. A climbing orchid with showy yellow red-spotted petals and red-flushed lateral sepals. Distributed in Assam region, (also in Indo-China); rare occurrence of ornamental value.

78. *Rheum nobile* Hook. f. & Thoms (Polygonaceae) Perenial herb. The "rhubarb" of eastern Himalayas (400-4500 m); rare; exploited for its leaves used as substitute for tobacco.

79. *Rhododendron arizelum* Balf & Forr. (Ericaceae)—Tree with highly ornamental rose-pink flower bunches. Discovered as new to India for Subansiri district, Arunachal Pradesh; rare occurrence; not reported from other parts of the eastern Himalayas.

80. *Rhododendron dalhousiae* Hook f. (Ericaceae)—A shrub by rhododendron with attractive large flowers; occurs in Eastern Himalayas; rare occurrence; exploited for horticultural purpose.

81. *Rhododendron edgeworthii* Hook f. (Ericaceae)— An Epiphytic ornamental shrub, reported from

Arunachal Pradesh; rarely collected; exploited for horticulture value.

82. *Rhododendron nivale* Hook. f. (Ericaceae)—Epiphytic herb. The smallest of rhododendrons only 5 cm. high, distributed in eastern Himalayas at an altitude of 5000 m; rarely collected; botanical interest.

83. *Rhododendron nutallii* Booth (Ericaceae)—Tree or epiphytic shrub; flowers large white with creamy base, sweet scented. Distributed in Bhutan and Daphla Hills, Arunachal Pradesh; rarely collected.

84. *Rhododendron Santapaui* Sastry et. al. (Ericaceae)—Pretty epiphytic shrub with white starry flowers. Collected so far only from Subansiri district, Arunachal Pradesh; very rare; endangered due to clearing of forests.

85. *Rhododendron stenaulum* Balf. & W.W.Sm. (Ericaceae)—Tree. Flowers white, fragrant, Discovered as new to India from Apathani valley, Subansiri District, Arunachal Pradesh. Only one tree was seen; rare occurrence; threatened due to the clearing of forests in the Valley.

86. *Rhus hookeri* Sahni & Bahadur (Anacardiaceae)—Small Tree. Endemic in Eastern India; rare occurrence.

87. *Sapria himalayana* Griff. (Rafflesiaceae)—A root parasite of botanical interest flowers ca 15 cm across. First collected in 1836 from Mishmi Hills, then collected after a century from Aka Hills, recently collected form Lohit district; distributed only in Arunachal Pradesh; very rare; of considerable botanical interest.

88. *Saussurea bracteata* Decne (Asteraceae)—Herb. A delicately fragrant species of western Himalayas; rare occurrence.

89. *Saussurea gnaphalodes* (Royle) Sch-Bip (Asteraceae)—A curious woolly herb of high altitude slopes of western Himalayas; rare occurrence.

90. *Saussurea lappa* C.B.C.I. (Asteraceae)—Herb. Endemic to Kashmir and Himachal Pradesh. Extensively exploited from the wild in the past; scarce and needs protection in the wild. The 'Kuth' of commerce now under cultivation.

91. *Schizaea digitata* Sw. (Schizaeaceae)—A perennial fern with discontinuous distribution in Khasi Hills, Sri Lanka and Malaysia; rare occurrence; of botanical interest.

92. *Tetracentron sinense* Oliv. var himalense Hara & Kanai (Tetracenttraceae)—A primitive vesselless dicotyledonous plant. Distributed in eastern Nepal, Bhutan and Arunachal Pradesh; rare occurrence; of considerable botanical interest.

93. *Thylacospermum rupifragrum* Schrenk (Caryophyllaceae)—Herb. Montypic cushion-forming plant of alpine Himalaya; rare; of botanical interest.

94. *Vanda coerulea* Griff (Orchidaceae)—Epiphytic. 'Blue vanda' of Asia, most beautiful of vandas. Distributed in Khasi & Jaintia Hills and Naga Hills; recently discovered in Tirap district, Arunachal Pradesh. Excessively exploited from the wild for ornamental purpose; now very scarce in the wild.

95. *Vanda pumila* Hook. f. (Orchidaceae)—Epiphytic herb. Ocassionally found in Sikkim and Bhutan; rare in the W. Himalaya.

96. *Vanilla pilifera* Holtt. (Orchidaceae)—Climbing orchid with fleshy leaves and creamy flowers. Garampani Sanctuary (Assam); very rare; also distributed in Malaya.

97. *Viola Falconeri* Hook. f. & Thoms (Violaceae)—A pretty herb. Endemic in Kashmir; rare occurrence.

98. *Zanthoxylum scandens* Bl. (Rutaceae)—Straggling shrub, distributed in Khasi Hills; rare occurence.

Rajasthan and Gujarat

99. *Commiphora wightii* (Arn.) Bhandari (Burseraceae)—Small tree. Exploited in the past for its gumresin from the ark. Distributed in Rajasthan and Gujarat in India; reportedly becoming scarce due to over-exploitation; presently cultivated.

100. *Helichrysum cutchicum* (C.B.CI) Rao et Deshpande (Asteraceae)—herb. Endemic in Gujarat; rare occurrence.

101. *Hyphaene dichotoma* (Baker) Furtado (H. indica Becc.) (Arecaceae)— The Indian doum palm curiouly branching. Distributed only in W. coast regions; of botanical interest.

102. *Rosa involluctrata Roxb.* (R Lyellii Linl.3(Rosanceae)— Subscandent shrub. Widely distributed in India; reportedly becominng scarce at Mt. Abu in Rajasthan due to clearing of forests.

Gangetic Plain

103. *Aldrovanda vesiculosa* Linn.(Droseraceae)—Aquatic herb, probably monotypic, carnivorous plant of botanical interest. Reported from salt pans near Calcutta; recently reported from Manipur and Bangladesh. Its habitat near Calcutta (Salt Lake area) is threatened due to reclamation.

Peninsular India

104. *Aparma barberi* Gamble (Aristolochiaceae)—Shrub. Endemic in Tirunelveli Hills; rare occurrence.

105. *Bentinkia condapanna* Borty ex. Roxb (Arecaceae)—Aead-like palm. Endemic in Kerala and Tirunelveli Hills; of botanical interest; rare occurrence.

106. *Ceropegia fantastica* Sedgw (Asclepiadaceae)— Slender climber described by sedgwick in 1921, based on a single collection from N. Kanark, and known only form Holotype (kew) until 1963, when this species was again collected in Goa (BSIWC);very rare.

107. *Cycas beddomei* Dyer (Cycadaceae)— Palm like tree. Endemic to Cuddapah Hills, S. India; rare.

108. *Dioscorea wightii* Hook f. (Dioscoreaceae)— Twinning climber with tuberous roots. Endemic in Tirunelveli Hills and recently collected after 100 years from type locality; very rare.

109. *Entada pursaetha* DC.(Mimosaceae)— Large woody climber with gigantic pads of botanical interest. Distributed in Himalayas. Assam and W. Peninsula; becoming rare due to large scale clearing of forests.

110. *Frerea indica* Delz. (Asclepiadaceae) undershrub. Endemic in W. Ghats S. India, rare occurrence.

111. *Gnetum ula Brongn.* (Gnetacea)— Extensive woody climbing gymnospoerm of botanical interest. Chiefly distributed in W. Ghats; Becoming rare due to large scale clearing of forests.

112. *Hoya-wightii* Hook. f. (Asclepiadaceae)— Large perennial climber, often becoming totally epiphytic Endemic to W. Peninsula becoming scarce due to clearing of forests and habitat destruction.

113. *Lilium neilgherrense* Wt. (Liliaceae)— The Nilgirilily. Endemic to the Nilgiris and a few high mountain ranges in S. India; rare occurrence.

114. *Lobelia nicotianaefolia* Roth ex Roem. & Schult.(*L. excelsa Leschen*) (Lobeliaceae)— Herb, distributed in W. Ghats; rare occurrence.

115. *Loeseneriella bourdillonii* (Gamble) Ramam.

(*Hippocrated bourdillonii* Gamble) (Hippocrateacea) — Climbing shrub. Endemic to S.W. India;rare occurrence.

116. *Manisuris devergens* (Hack) Kuntz. (Poaceae)— Tufted grass. Endemic to peninsular India; collected after 130 years from Coorg district; rare occurrence. All the 12 species of *Manisuris* found in India are endemic; ten of them confined to South India.

117. *Piper barberi* Gamble(Piperaceae)— Climbing shrub with pinnately-veined leaves; endemic in Tirunelveli Hills; botanical interest; rare occurrence.

118. *Podocarpus wallichianus* (Podocarpaceae)—Tree. The only native conifer of peninsular India, also occurs in E. Himalayas Khasi Hills and Nicobar Islands; rare occurrence.

119. *Pterocarpus santalinus* Linn. f. (Papilionaceae)— Tree. Endemic in Cuddapah district, S. India; the Red Sanders of Commerce, exploited much in the past for its beautiful red wavygrained wood. Export already banned.

120. *Pterospermum obtusifolium* Wt. (Sterculiaceae)— Tree. Distributed in Malabar and Karnataka; rare occurrence.

121. *Santalum album* Linn. (Santalaceae)— Tree. Usually semi-parasitic. Occurs in peninsular India. Sandlwood of commerce, much exploited in trade, becoming rare (now cultivated).

Andaman & Nicobar Islands

122. *Ailanthus kurzii* Prain (Simaroubaceae)— Tall tree. Endemic in Andaman Islands; becoming rare due to clearing of forests and not collected during the recent botanical explorations in these islands.

123. *Canarium mannii* King (Burseraceae)— Tree. Distributed in Middle Andaman Island. Parkinson reports it as not uncommon but this species has not collected during the recent botanical explorations; probably becoming rare.

124 *Dipterocarpus kerrii* King (Dipterocarpaceae)— Tall tree. Occurs in South Andaman. "Seen only in the Goplakbang Valley, where a sample plot of this has been made" (Parkinson, For Fl. Andaman Island, P. 93, 1923). This species has not been collected during the recent botanical explorations in Andaman Islands; rare.

125. *Hippocratea nicobarica* Kurz (Hippocateaceae)— Climbing shrub, distributed in tropical forests of eastern coast of Katchall Island (Nicobar group); rare occurrence; not collected during the recent botanical explorations.

126. *Lagerstroemia hypoleuca* Kurz (Lythraceae)— Tree. Widely distributed in Andamans. This species is one of the important timber yielding trees and has been exploited much; becoming scarce in wild.

127. *Myristica andamanica* Hook. f. (Nyristicaceae)— Tree. Distributed in middle Andaman, South Andaman and Porlob Islands; becoming scarce due to clearing of forests. Not collected in the recent botanical explorations.

128. *Ophioglossum pendulum* Linn. (Ophioglossaceae)— Epiphytic herb with long ribbon like fronds. Recently reported from Campbell Bay, great Nicobar Islands; rare occurrence of botanical interest.

129. *Podocarpus nerifolius* D. Don (Podocarpaceae)— Tree. The only confier of the Andamans, although widely distributed; exploited for its valuable timber; becoming scarce in these Islands.

130. *Psilotum complanatum* Sw. (Psilotaceae)—Epiphytic pendulous herb, recently reported from Great Nicobar Island; rare occurrence; of botanical interest.

131. *Symplocos odoratissima* (B1.) Choisy ex Zoll (*S. Chengapae* Raizada & Sahni) (Symplocaceae)— A shrub or tree. Distributed in Great Nicobar Island; rare occurrence. Also occurs in Indonesia, Malaya, Singapore and Philippines.

132. *Uvaria nicobarica* Raizada & Sahni (Annonaceae)—Woody climber, Endemic in Great Nicobar Island; rare occurrence.

Gaps in our Knowledge Rare and Endangered Indian Plants

The main lacunae in our knowledge with regard to this subject lies in the present actual status of the species i.e. the exact number and location of populations of these species occurring in the wild, quantities or the numbers collected for economic exploitation or other purposes; their capacity to reprocduce and propagate themselves in nature and their populations's or representatives in cultivation under artificial conditions in farms or experimental gardens.

Such observations can be made by repeated visits to the field and need concerted efforts of several botanists and constant monitoring. This naturally an enormous task for a country like India of the size of a sub-continent. A recent close study of the orchid genus coelogyne serves as a good example four-species of the genus viz C. Albo-lutea, C, treutleri, C. brevifolia and C. assamica, are know only by their mention in literature and types (specimens/ illustrations). These could not be located even in their type localities inspite of several recent botanical explorations and are believed to have become extinct. Excepting the types even their preserved specimen are not available in any herbarium. Only such intensive library field and herbarium studies can reveal the real status of our threatened taxa. Similarly the recent discovery of the Indian pitcher plant. Nepenthes khasiana in Garo Hills near Tura a locality other than its own type locality near Jowari also highlights the need for intensive botanical explorations and constant monitoring of the floras in the country.

Actions Taken to save the Rarities of Indian Flora

On the the recommendations of the Botanical Survey of India and other concerned agencies, the Government of India has imposed restrictions or a complete ban on the export of certain species. The Wildlife Wardens/Chief Conservations of forests in such State have been appointed as custodians of flora and fauna to check illegal trade practices in these species. They can be exported only when certified by the specified authority (Presently the Director, Forestry & Wildlife)

India is one of the signatories to the Convention on International Trade in Endangered Species of Wild Fauna and Flora (CITES, 1973), it has been ratified so far by 67 countries of the world, for strict regulation and control of trade. According to the Convention, if a species is even smuggled out of a country. The Country destination will co-operate by obstructing the import of the consignment.

The National committee on Environmental Planning and Coordination (NCEPC) and the National Committee on Man and Biosphere (MAB) have also been concerned with the protection habitats which have undisturbed natural vegetation several such areas have been identified for conservation and preservation as 'Biosphere Reserves' throughout the country. It is also proposed to set up Germ Plasm Banks and Gene Sanctuaries to protect the areas rich in wild relatives of certain economic and cultivated plants.

The Botanical Survey of India with its main Botanic Garden at Sibpur and the regional experimental gardens established at different altitudinal situations to suit the acclimatizations of plants occurring in different ecological conditions in the country, have in their holdings rare, endemic and threatened plants. Several species of orchids are also under cultivation and protection in the National orchidaria or Shillong, Yercaud and Calcutta. Of these, *Vanda Coerulea* (Blue Vanda) *Dendrobium nobile cymbidium grandiflorum, Renanthera* and species of *paphiopedilum* and pleione growing in Shillong deserve special mention.

The Government of Meghalaya State has taken action on the recommendations of the Botacnial Survey of India, and declared the *Nepenthes khasiana* habitats as protected areas to preserve the plants in *situ.*

Future Programme of Work in the Botanical Survey of India

In order to fill the gaps in our knowledge on threatened plants the following studies are being undertaken by the Survey.

1. A review of literature and study of herbarium specimens of endangered or rare species deposited in the various herbaria of the country to collect data on the distribution, abundance and frequency.
2. Fresh explorations to areas for collection of plants and data on rare/endangered species with regard to the present distribution frequency and phenological data.
3. Collection of live plants for introduction into the botanical gardens for protection and further studies on propagation.
4. Preparation of distribution maps of rare and endangered plants.
5. Preparation of an illustrated book on rare and endangered plants.

6. Preparation of an identification manual of plants banned for export to facilitate proper identification and for education of public.

As many as 500 million kinds of plant, animal and micro-organism have made this planet home since life began over 3.5 billion years ago. Today, there are only 5 million to 10 million species alive—we do not know exactly how many, because there are many biologically uncharted areas such as the tropical rain forests, where some estimate that such as the tropical rain forests, where some estimate that over 90% of the living organisms remain unclassified. Thus, since life began, about 490 million species have become extinct.

A Study of the history of life from an evolutionary perspective teaches us that biological extinction is a fact of life. As species evolve, the old give way to the new. Many of the so-called extinct species are represented today by their descendants. Thus, many species did not die off as such but were simply altered during evolution.

Other species vanished completely as a result of drastic climatic changes or increasing environmental resistance created by excess predation or disease. The dinosaurs, for example, perished rather abruptly all over the world 65 million years ago, possibly as a result of global cooling.

Much more recently, at the end of the pleistocene epoc 10,000 years ago, many large mammals such as the sabertooed tiger and giant tree sloth vanished from North America. Their extinction resulted either from climatic changes at the end of the Ice Age or possibly from Siberia. As these people moved south and east, they hunted the large mammals such as the wooly mammoth wooly rhinoceros giant deer, ground sloth, mastodons, giant beavers, sabertoothed tigers, bison, and elk. They undoubtedly also competed for the same food resources.

From the end of Pleistocene until now, the earth's climate has been fairly constant and relatively few plants and animals have become extinct in the 1800s. 84 animals species and subspecies are known two have vanished. By 1984 the number had risen to 184 and if trends continue, many scientists believe, 170 more species will disappear by 2000 bringing the total to more than 350 if conditions turn considerably worse. More than 700 additional species might by destroyed before the end of the century.

One vertebrate species is becoming extinct every nine months, compared with a natural rate of one species every thousand years. When plants insects and microbes are added, the rate may be as high as one species per day. Ecologists argue that plant extinctions will have much more profound impacts on the ecosystem that the extinction of animals because plants form the base of the food web.

Some authors assert that if the world's population continues to grow and nations continue to destroy wildlife habitat by expanding agriculture and other activities at the current pace, the extinction rate could reach 20,000 species a year. At this rate 400,000 species of plants and animals could be wiped out in 20 years. World wildlife expert Norman Myers argues that we could lose 1 million species by 2000. Although such high rates of extinction are probably exaggerated extinction will undoubtedly increase unless we make conscious efforts to curb population growth and properly manage our renewable and non-renewable recources.

Why Save Endangered Species?

Wildlife expert Norman Myers once wrote: "We can marvel at the colours of a butterfly the grace of giraffe, the power of an elephant of the delicate structure of a diatom. Every time a species goes extinct, we are irreversibly impoverished." Aesthetics in itself is a good argument for saving species. As we preserve are collection for future generations, so we should preserve this living art form that painters and sculpture only dream of recreating.

A second reason for protecting species is that other organisms have a right to live. Can we not guarantee them the right to exist, something that we ourselves doggedly strive for?

The third reason for preserving other organisms is a matter of ethical responsibility. As a species capable of destroying the world, some ecologists and philosophers feel, we have a moral obligation to preserve organisms with lesser powers.

The fourth reason for protection is that it makes good economic sense to do so. As Myers wrote : "From morning coffee to evening nightcap we benefit in our daily life style from the fellow species that share our One Earth home. Without knowing it we utilize hundreds of products each day that owe their origin to wild animals and plants. Indeed our welfare is intimately tied up with the welfare of wildlife. We may conservationists proclaim that by saving the lives of wild species, we may be saving our own."

From the biosphere we reap countless benefits : fish from the sea, medicines and other products from plant important plant and animal genes needed to improve domestic crops and livestock, something for hunters to shoot at something for anglers to hook and reel in some thing for bird-watchers to peer and research animals that help provide valuable insights into human physiology and behaviour.

The economic benefits of these activites can be enormous. In 1983 alone trout and salmon anglers spent an estimated $5 billion on equipment transportation food and lodging and hunters and bird watchers spent billions on tennis shoes, supplies, books gasoline food and lodging. Billions of dollars pour into companies that cut timber or extract valuable medicines from plants.

Each time we take a prescription drug, the chance are one in two that its origin was a wild plant. The commercial value of such drugs is around $15 billion per year in the United States and about $40 billion world wide.

The U.S. Department of Agriculture estimates that each year genes bred into commercial crops yield over $1 billion worth of food. Similar gains can be documented for Canada, Great Britain, the Soviet Union, Australia and other agricultural nations. About half of the increased productivity in corn over the past 50 years has resulted from "genetic transfusions" from wild relatives of corn or primitive cultivars in use only in isolated regions. Some biologists argue that the current high-yielding corn, which produces 250 bushels per hectare (650 bushels per acre), could be increased fivefold with further genetic transfusions from this ancestors.

New developments are on the horizon that may offer financial gains and a more healthful life for us. For example the adhesive that barnacles use to adhere to the bottoms of ships may provide us with a substance to cement tooth fillings in place. A chemical derived from the skeletons of shrimps crabs and lobsters (chitosanase) may serve as a preventive medicine against fungi infections. A few drug that kills viruses may be developed from a Caribbean sponge which till now physicians have been virtually helpless in the battle against viruses.

The future of the biological world is one of grave

economic concern. The loss of fish from acid, rain and over fishing results in a loss of income to owners of fishing lodges and sporting goods stores. Ecological impoverishment will keep more of us at home, diminishing income to the recreation and tourism industry.

Plants and animals lost before they are explored for possible benefits will diminish our opportunities to fight disease and increase productivity. a dozen genetic reservoirs (so-called centers of diversity) located in the tropics and subtropics are the source of virtually all commercially value plants and animals. And they provide genetic material needed in the continual battle to improve plant and animal resistance. Loss of these centers would have a global impact on food supplies.

The final reason for preserving species is to protect ecosystem stability. All living organisms are an integral part of the biosphere and provide invaluable services. These include thte control of pests, the recycling of nutrients, the replenishment of atmospheric oxygen, the maintenance of local climate, the control of groundwater levels, the regulation of atmospheric carbon dioxide and global climate and the control of floding. It is no exaggeration to say that civilization depends on the survival of the biological world. Perhaps in the short term a species lost here and there may be of little consequence for ovirall ecosystem stability but in the long term the cumulative effect of such losses may someday threaten our existence.

Critics of the animal protection movement argue that too much effort is directed to the saving of endangered species. Why was there so much fuss over the tiny snail arter a fish imperiled by the Tellico Dam built by the Tennesse Valley Authority (TVA). Surely this one species can have no great significance to us although no irreparable ecosystem damage would result from its extinction, the question that arises is where to draw the line. Wildlife advocates argues that if we take the attitude that each species by itself is dispensable bit by bit we will destroy the righ biological world we live in. Somewhere the line has to be drawn each endangered species is worth saving because it stops the momentum toward widespread destruction. In growth-oriented societies this momentum may be diffecult to stop, much less slow down. Therefore each hurdle put in its way becomes an important force in saving the living creatures that make up our web of life.

At one time our wild life was as remarkable for its abundance as it is even now for its variety, with tigers hunting the forests in hundreds and antelopes the open plains in thousnands upon thousands. Over the centuries it has played an important role in our economy and culture—as a source of protein for the masses, as items or trade, and as an inspiration for a rich religious lore. The wild life has gradually declined over the millenia and very rapidly since the British took over the country. At last we have reached a critical stage at which most, if not all of our wild life appears on the verge of being totally wiped out. Only bold and imaginative action in conservation and management will enable us to save this rich heritage from an irreversible loss at this critical juncture.

The Indian subcontinent may be divided into the following eight regions for our purpose.

(1) Punjab and the adjoining semi-arid and desert regions.

(2) The alluvial plains of Gangas and lower Brahmaputra

(3) The western peninsular tract excluding the Western Ghats

(4) The eastern peninsular tract
(5) The Western Ghats or Sahyadris
(6) Ladakh and adjacent regions
(7) The forested hill-slopes of Himalayas
(8) Assam and adjacent hill tracts.

Each of these ecological regions has been subjected to distinctive demagogic pressures commercial exploitation, agricultural expansion and other modern deve lopment activities, which are so much a part of the present day industrial and agricultural economy.

Today the position of wild animals is in general alarming and many species are greatly endangered and are fighting a losing battle for survival. Happily a general awareness has developed among the informed public about the necessity to conserve the natural ecosystems of the country and the living wild communities of animals and plants these contain. To ensure this, certain legal steps to regulate exploitation of wildlife have been taken by the Central and state Governments. Endangered and threatened species of animals are totally protected and barring vermin of very common species, exploitation of the rest is regulated by law under the provision of the Wildlife (Protection) Act, 1972.

To gain a full appreciation of the endangered fauna of the country, a brief account of the occurrence of different species of Mammals, Birds, Reptiles, Amphibians, and Insects in the country and their endangered forms, is presented below.

Mammals

About 850 species constitute the mammalian fauna of India of which 81 are greatly endangered.

The order primates are represented by 19 species, of which small and slender-bodies, and is restricted to the tropical rain forests and open woodland forests of South India while the slow loris (Nycticebus coucang) is slightly bigger and inhabits the denser forests of north-east India. They are persecuted chiefly for their flesh and for the supposed medicinal value of various parts of the body. Among the macaques the liontailed (Macaca silenus) is one of the world's most endangered primates; surviving in the evergreen forests of the Western Ghats of South India, its total population is estimated to be about 800 only the pigtailed (*M. assamensis*) occur in the dense evergreen forests of the north-eastern India, while the crabeating (*M. fascicularis*) is restricted to a few islands in the Nicobar group. All of them are exploited for economic purposes and the first two species for their fesh and for alleged medicinal value. In the langur group, the Nilgiri langur (*Presbvtis johni*) is a multihabitat species occurring in addition to the sholas, in the temperature evergreen forests above c. 1700 mt. altitude in the Western Ghats. Its status is not as precarious as that of the liontailed macaque, it is hunted for its flesh or for medicinal properties, the skin used for making drums, its fur is also prized. The cappd (*P. pileutus*), the golden (*P.geei*) and Phayre's leafmonkey (*P.phayrei*) inhabit tropical and deciduous dense forests in the north-eastern India. The only ape found in India is the Hoolock gibbon (*Hylobaes hoolock*), which thrives in the hily forests of north-eastern India. It is hunted by local tribes for its flesh and trapped for sale.

Of the family Ursidae, the Himalayan brown, Bear (*Ursus arctos*) occurs in the upper reaches of the Himalaya from Kashmir east to Sikkim, occasionally coming down to lower reaches in severe winter the Malayan sun Bear (*Helarctos malayanus*) inhabits the hilly terrains of northeastern India, south of the Brahmaputra and the sloth Bear (*Melursus ursinus*) in the well-wooded forests all over India. They are hunted for their fur and gall bladder which is supposed to have medicinal properties, also for flesh fat and other derivatives.

The family Mustelidae has 14 species in India, of which the Ermine (*Mustelidae erminea*) occurs in the western Himalaya and is hunted for its pelt; while the ratel (*Mellivora capensis*), is found all over the country and is killed mainly to protect poultry from it and to some extent for its skin.

The family Viverridae is represented by 15 species of which the Malabar Civet (*Viverra megaspila*) occurs in the coastal districts of Kerala and is feared to have been nearly exterminated; the spotted linsang or tiger-civet (*Prionodon pardicolor*) rangers from Nepal east to Arunachal Pradesh extending further eastword in the mountain and hill forests, but is nowwhere common.

The striped hyaena (*Hyaena hyaena*) is common in the open country with low hills and ravines throughout India. It is prosecuted for its habit of predating on livestock and occasional childlifting.

There are 15 Indian species in the family Felidae, of which the tiger (*Panthera tigris*) is a multihabitat species found practicaly throughout the country except the desert region even upto c. 3050 mt. height in the Himalaya. Unrestricted killing for trophies, skins for commercial purposes, flesh and fat for medicinal use, derivatives of body parts of their supposed magical properties, and sometimes through eradication of proclaimed animals as man-eater or Cattle-lifter, had declined its population to 1827. Through 'Project Tiger', its population has now increased to 2278 within a few years. The Indian lion (*P. leo*) is restricted to the Gir forest in Gujarat. The leopard (*P.Pardus*) and the snow leopard (*P.uncia*) are two other big cats. The former is widely distributed throughout the country in wooded forests, while the latter is restricted to higher reaches all along the Himalaya, from 2000 to 4000 mt. altitude. These have been largely hunted for commercial purposes or for the protection of livestock. Among the lesser cats, the desert cat (*Felis silverstris*), the lynx (*F. lynx*), and the caracal (*F.caracal*) are found in the arid and semi-arid tracts of the western and north-western India while the jungle cat (*F.chaus*) and the leopard cat (*F. bengalensis*) are found throughout the country in small numbers. Many species of cats are hunted for their pelt. The cheetah or hunting leopard (*Acinonyx jubatus*) which once ocurred in the planis and the hilly regions of northern central and peninsular India south to Karnataka, is already extinct from India.

The dugong (*Dugong dugong*) among the Sirenia, occurs in the Gulf of Kutch, on the coast of Malabar, Gulf of Manar and the Andaman and Nicobar Islands. Once an important dugong 'fishery' existed in the Gulf of Manaar for its flesh, but the species are now scarce there.

Among the Perissodactyla, the Great Indian one-horned rhinoceros (*Rhinoceros unicornis*) as once extensively distributed in the Gangetic plains, but is now restricted in India to North Bengal and the plains of Assam. It was on the verge of extinction in the beginning of this century when 1904, only about a dozen individuals remained in Kaziranga (Assam) and fewer in North Bengal. It was in great demand for the alleged aphrodisiac properties of its horn and almost every part of the animal including flesh was in great demand. Though the most sought after part of the rhino was its horn, its urine was also credited with antiseptic properties. It became the victim of unrestricted killing and trapping. Measures for conservation adopted since 1904 have helped to rehabilitate it, and today its population is estimated to be around 1000 in the Indian territory. The Indian wild ass (*Asinus hemionous*) is restricted to the saline flats of the Rann of Kutch. Its population has dwindled greatly form 5000 in 1946 to 870 in 1962. The present population is estimated to be 720. The major reasons for its decline are probably Surra disease and regular disturbance in their habit at by the saltworkers.

In the family Cervidae, out of the nine species; five are endangered. These are the Kashmir stag or hangul (*Cervus elaphus hanglu*) which is restricted to the northern side of the Kashmir Valley and some adjacent valleys, its population is estimated to be about 100 in Dachigam near Srinagar and another 200 elsewhere in Dessu in the Bringi Valley and Chumki, overa and Khiram in the Lidder Valley. The swamp deer or barasinagha (*C. duvaucelli*) is represented by two sub-species, the swamp-dwelling (*C.d. duvaucelli*) which lives in, marshy tracts of the Tarai and Duars from northern parts of the upper Gangetic plain east to Assam. The other, the hard ground living (*C.d. branderi*) thrives on the open grassy land of Madhya Pradesh. The population of both the sub-species together in critically low now mostly confined to the Kanha of the Himalaya from Chamba Valley in the west to Sikkim in the east and the forest musk deer (*M. chrysogaster*) is found from Garhwal in the west to Arunachal Pradesh and Assam in the east. The family Bovidae contains 21 species of which the under mentioned 14 are endangered. The blackbuck or Indian antelope (*Anteope cervicapra*) occupies open plains, while the chinkara or Indian gazelle (*Gazella dorcas bennetii*) is basiclly an animal or the desert, but it also inhabits the grassy plans and scrub zones of the country. The black buck population is about 10,000 to 15,000 in the western and southern India, while the chinkara is over 500 in Gujarat alone. The chiur or Tibetan antelope (*Pantholaps hodgni*) and the Tibetan gazelle (*Procapra picticaudata*) are both found in the fill forests of India. The wild buffalo (*Bubalus bubais*) is now restricted to certain swampy areas of Assam and Madhya Pradesh its total population around 1000, being 500 in Kairanga 300 to 400 in Manas (both in Assam) and about 100 in Bastar (Madhya Pradesh). The urial or shapu (*Ovis orientilis*) the Markhor (*Cara falconeri*) and the Himalayan ibex (*C. ibex*) frequent the grassy patches in the high altitude regions of Kashmir, Himachal Pradesh and Punjab, and the nayan of Great Tibetan sheep (*Ovis ammon hodgsoni*) in the plateau region from Ladakh east to Sikkim on the higher reaches of the Himalaya. The serow (*Capricornis sumatraensis*) occurs in the thickly wooded areas of the Himalaya between c. 1850 to 3050 mt. altitude from Kashmir east to Arunachal Pradesh.

The Lagomorphe represented by four species, of which the hispid hare or Assam rabbit (*caaprologus hispidus*) is a are and elusive species found in the grassland of the Tarai an the Duars and also in the Brahmaputra Valley.

The order Rodentia forms the largest group represented by 95 species within the Indian territorial limits, of which 11 species of flying squirrels and two of marmots are endangered, being widely exploited for their pelt as well as for their flesh. The common giant flying squirrel (*Petaurista petaurista*) is widely distributed throughout the wooded parts of the country but occurs in small numbers. The wooly flying squirrel (*Eupetaurus cinereus*) is localised in the high altitude regions of Kashmir and Sikkim. The Smaller Kashmir flying squirrel (*Hylopetes fimbriatus*) is confined to Kashmir. The Himalayan marmot (marmota bobak) occurs in the high altitudes of the Himalayas from Kashmir east to Sikkim, while the Long-tailed marmot (M. caudata) is restricted to the high altitudes of Kashmir within the Indian limit.

As many as 25 species of Cetacea are known from the Indian territorial limits and are all endangered. The

Gangetic dolphin (*Plastanist gagetica*) is found in the river system of the Ganga and Brahmaputra. Other cetaceans are marine forms. The baleen whales (*Balaenoptea spp; Megaptera sp.*) and other whales and dolphin species (*Kogia sp; Stenella spp; Sotalia spp; Steno sp; Tursiops spp; Lagenorhynchus sp; Ziphius sp; Neomris sp; and Delphinus spp.*) have been indiscriminately exploited for ballen for commerce, and for flesh, fat, bone and ambergris.

Birds

There are about 2100 species and subspecies of birds known in India, of which a large number of birds are greatly endangered. Below is a brief account of these species.

Of the ducks, geese and swans (*family Anatidae*), the pinkheued duck (*Rhodonessa caryophyllacea*) occurred in forest ponds and swamps along he Himalayan foothills and Terai, from Nepal east of Manipur and also appeared sporadicaly in places as far away as Punjab in the north and Tamil Nadu in the south. At present, it is probably extinct from India. The whitewinged wood duck (*Cairina scutalata*) breeds on trees in dense swampy forests of Assam, Arunachal Pradesh and Manipur. Migratory swans, mely, the whistlingswan (*Cygnus columbianus*), the whooper swan (*C. cygmus*) and the mute swan (*C. olor*) are exceedingly rare vagrants or stragglers. They are disappearing for want of appropriate habitat.

Most of the diurnal birds of prey belonging to the families Accipitridae. Pandionidae and Falconidae live a very precarious existence due to habitat destruction, pollution hazards, poison contamination, trapping for falconry and are subject to random killing for their tendency to prey on poultry etc. Vaious species of hawks (*Accipiter spp.*), eagles (*Aquila spp.*) and falcons (*Falco spp.*) are widely distributed but in small numbers, and they are trapped for falconry. The whitebellied sea eagle

(*Heliaeetus leucogaster*) of the Indian coastal areas and the Bay Islands are of small population size.

The Nicobar megapode (*Megapodius freycinet*), family Megapodidae is confined to the Nicobar group of islands within the Indian limits. Its previous abundance has been affected by habitat destruction, increased human interference and exploitation, specially for its flesh and eggs.

Phasianid birds (*family Phasianide*) are in great danger chiefly from human exploitation for their flesh they have always been branded as 'game birds'. Many of these are gorgeously coloured (*Ornamental*) and are in heavy demand from zoos and menageries for which they are trapped.

Bamboo partridge (*Bambusicola futchi*) restricted to open scrub and bamboo jungles from the plains to about 2000 m altitude in north-eastern India is fast loosing its habitat due to human interference. The mountainquail (*ophrysia superciliosa*) known only from a few specimens taken in the late 19th century in Kumaon is now thought to be extinct. The blood pheasant (*Ithaginis cruentus*) occurring in the Himalayas between 2000 nd 4500 m altitude from Nepal east to Arunachal Pradesh, the western Tragopan (*Tragopan melancephala*) of the western Himalayas from Kashmir to Garhwal between 1000 and 3000 m altitude of the Himalaya from eastern Garhwal east to Arunachal Pradesh between 2500 and 4000 m altitude. The Indian peafowl (*Pavo cristatus*), although not threatened, needs to be mentioned since it is the National Bird of India.

Of the cranes (*family Gruidae*), the eastern common crane (*Grus grus Lilfordi*) is a winter migrant in the larger wetlands of northern India, the blacknecked crane (*Grus nigricollis*), a greatly threatened bird, breads in Ladakh at the altitude above 3000 m and presumably also in the eastern Himalayas, the Hooded Crane (*Grus monacha*) is a rare winter migrant to eastern Assam, the Siberian crane or the great white crane (*Grus leucogeranus*) is a regular winter visitor to the protected wetlands of northern Indian plains but a decidedly diminishing species.

The masked finfoot (Heliopais parsonata, family Heliornithidae) a unique relic species, is known from eastern Assam and Arunachal Pradesh in dense lowland jungle swamps and ponds. It is so rate that no authentic record of its occurrence is available for the last fifty years or so.

Of the bustards and floricans (*family Otididea*) the little bustard (*Otis tetrx*) and the houbra or Macqueen's bustard (*Chlamydtis undulata*) are winter migrants in northern and north western India; the great Indian bustard (*Ardeotis nigriceps*), the Bengal florican (*Eupodotis bengalensis*) (*Sypheotides indica*) ae resident species. Their previous abundance as well as range have shrunk greatly and the birds are facing extinction.

Of the pigeons and doves (*family Columbidae*), the Nicobar pigeon (*Caloenus nicobarica*), restricted to the Nicobar group of islands, is fast loosing its breeding grounds and is confronted with slaughter by the local people.

The hornbills (*family Bucerotidae*) are large birds, and most species are threatened, chiefly due to deforestation resulting in a dearth of suitable nesting trees, and persecution by man for alleged medicinal value of their flesh, fat and bone, and for their casque and other derivatives used as ornaments, charm or in witchcraft by the tribal people. The most threatened of them are : the whitethroated brown hornbill (*Ptilolaemus tickelli*) found in the hills of eastern India in evergreen and curover scrub forests from 300 to 1500 m altitude; the rfusnecked hornbill (*Aceros ripalensis*) found in central and eastern

Himalayas in dense deciduous and evergreen forests of Sikkim, Arunachal Pradesh, Manipur and Mizoram; the wreathed hornbill (*Rhyticeros undulatus*) found in the tropical and subtropical evergreen forests along the Himalayan foothills in the Duars and also in the region south of Brahmaputra from the plains to an altitude of 2500 m; the Narcondam Bornbill, (*Rhyticeros (undultus) narcondami*) an endemic species in the Narcondam Island of the Andaman group is highly vulnerable on this volcanic island; the great pied hornbill (*Buceros bicornis*) of the tropical and subtropical evergreen, and most deciduous forests from the plains up to 2000 m altitude, from Kumaon east to north-eastern states, and in the Western Ghats from Khandala (Maharashtra) to Kerala; the Indian pied hornbill (*Anthraceoceros malabaricus*) of moist deciduous and evergreen forests and mango groves from the plains to 350 m altitude along the terai from Punjab east to north-eastern States, and in Bihar, Orissa, Andhra Pradesh and Madhya Pradesh; an the Malabar pied hornbill (*Anthrceceros Coronatus*) of evergreen and moist deciuous forests from the plains to 300 m altitude from Ratnagiri (Maharashtra) east to Orissa and South to Karnataka, Tamil Nadu and Kerala.

Reptiles

The Reptiles of India belong to the following groups; tortoises and turtles (*Chelonia*), crocodiles and gharials (*Loricata*), lizards (*Sauria*) and snakes (*Serpentes*). Exploitation of reptilian resources in India is chiefly for flesh and eggs as food, hide, skin and shell for fancy leather articles and handicrafts, teeth and bones as charms, fat for alleged medicinal properties and venom for preparation of medicinal products. Living snake specimens are also utilized by snake charmers and there is a great demand from the zoos and menageries.

The flesh of turtle, tortoise and terrapin is widely

consumed almost throughout India. They are collected from both inland and marine sources. Indiscriminate capture for trade in their flesh and shell, and collection of eggs from rookeries have resulted in many species becoming rate. Particularly endangered are the leatherback or trunk turtle (Dermochelys coriacea), the green sea turtle (Chelonia mydas), *the loggerhead or olive ridely* (Lapideoheyls livacea) and the hawks bill or tortoiseshell turtle (Eretmochelys imbricata). They are all marine forms and found in the coastal areas and estuaries. All the three crocodilians found in India, viz. the estuarine crocodile (Crocodylus porosus) found the estuaries of larger rivers and coastal areas, including those of the Bay Islands, the marsh crocodile (Crocodylus palustris) occurring in inland swamps and larger waterbodies all over India, and the gharial (Gavialis gangeticus) found in the big rivers with islands littered with gravel, boulders and pebbles, all lead a precarious life.

All the five species of monitor lizards, namely the common Indian monitor (Varanus bengalensis) widely distributed throughout India, the desert monitor (Varanus griseus) found in the arid and semiarid zones of western, north-western and west-central India, the yellow monitor (Varanus falvescens) found in northern India in the Ganga Valley, the water monitor (Varanus salvator) found in the estuarine and freshwater swampy areas of eastern and south-eastern part of the country, and the clouded monitor (Varanus nebulosus) found in eastern India and the Bay Islands are greatly threatnened.

Of the snakes, the Indian or rock python (Python molurus) found in the dense forested tracts throughout India and the reticulated python (Python reticulat) restricted to tropical and subtropical forests of eastern and north eastern Indian and in the Bay Islands are largely persecuted for their ornamental skins.

Of the amphibians, the Indian Salamander (Tylototriton verrucosus) is of much zoological interest and is found in eastern and north-eastern India. It is restricted to some small pockets in the Darjeeling District of West Bengal, Sikkim, Arunachal Pradesh and Manipur. Habitat destruction and aquatic pollution, it is fered, may greatly endanger its survival.

3 Ethnobotanical Studies in India

The history of medicinal plants is intimately connected with the history of plants. Primitive man lived at the mercy of nature, in constant terror of diseases. From the earliest times, tribal priests and medicine men (with doctors) used various plants, minerals and animal organs usually in association with strange rituals and incantation, to drive out the evil spirits which they believed to be the cause of the disease. Astonishingly, these magical rites seemed to help. In some primitive tribes, a victim of disease was half-buried in soil for several days to exercise the malevolent spirits which has possessed him. Among the extremes of treatments was the chipping of holes in the skull to release the tormenting evil spirits. This theory of demoniacal possession lasted many centuries and exists even today in areas where people still live in primitive societies.

Records of early civilisation in all parts of the world reveal that a considerable number of drugs used in modern medicine were in use even in ancient times. The use of plants for curing various human ailments figured in ancient manuscripts such as The Bible, The Rig-Vedas, The Iliad and The Odyssey and the History of Herodotus. Over 6000 years ago, the ancient Chinese were using drug plants. The Egyptians, Babylonians, Sumerians, Greeks and Romans, all developed their respective characteristic Materia Madica. On the other side of the world, the Aztecs, Mayans and Incas had all developed primitive medicine. Some of the ancient Egyptian text-books 'papyri' (such as the Edwin Smith Papyrus and the Ebers Papyrus), written as early as 1600 B.C., indicate that the Egyptians had an amazingly complex Materia Medica. Apart from the names of many medicinal plants then known the papyri also included several hundred recipies or prescriptions for various diseases. The Edwin Smith Papyrus is now one of the prized collections of the New York Academy of Medicine.

In India the ayurvedic system of medicine has been in use for over three thousand years. Charaka and Susruta, two of the earliest Indian authors had sufficient knowledge of the properties of the Indian medicinal plants. Their medical works the Charaka Samhita and the Susruta Samhita are esteemed even today as treasures of literature on indigenous medicine.

The Greeks and Romans were familiar with many of the present day drugs as is evident from the work of Hippocrates, Aristotle Theophrastus, Pliny the Elder, Disoscrides and Galan. They wrote extensively about medicinal herbs giving their names along with a description of each plant, illustrations, their putative healing properties and also complex descriptions for the preparation of medicines. Hippocrates, the 'Father of Medicine'was the first to attempt a scientific explanation for diseases. His influence remains today in the hippocratic oath taken by young doctors upon their graduation from medical school. Dioscorides treatise on medicinal plants De Materia Medica remained the supreme authority for over sixteen centuries, during which the manuscript was laboriously copied and recopied with few additions. In like manner, the works of Aristotle, Galen and even Pliny were of Aristotle, Galen and even Pliny were copied and handed down with increasing inaccuracies. During the Dark Ages (A.D. 400-1000) few new ideas were added. During the Middle Ages (A.D. 1000-1500) also little significant botanical progress was made. Had biology progressed steadily from the time of Aristotle, there probably would have been no great plague and small pox epidemics during the Middle Ages. One epidemic of the plague between 1347 and 1350 killed sixty million people in Europe, Asia and Africa.

About the beginning of sixteenth century, several herbals of considerable merit were published such as those of Brunfels (1530), Bock (1539), Fuchs (1542), Cordus (1561), and L'Obel (1576). Although these works were a great improvement over earlier ones, they still propagated many myths and superstitions such as the Doctrine of Signatures' advocated by an eccentric genius Paracelsus (1493-1541), according to which all plants possessed certain signs given by God, which indicated their usefulness in treating diseases of similarly shaped organs in human body. Plants, for example, with heart-shaped leaves were used for cardiac disorders; the sap of bloodroot (Sanguinaria canadensis L.) as a blood tonic; the walnut with numerous invaginations and convolutions for brain diseases and pomegranate seeds for dental diseases. Particularly striking was the myth of the mandrake plant (Mandragora officinarnum L.) of the family Solanaceae. The fleshy and often forked roots of this plant somewhat resemble the torso and the lower limbs of a human figure and were believed to be quite efficacious for creating various human ailments. Since ancient Greek times, the resulting decoction of the mandrake roots either prepared in boiling water or in wine, has had a reputation of producing sleep. It was perhaps the first true anaesthesia and later in 1889 was found to contain a mixture of paindeadening alkaloids (podophyllin, mandragorin and hyoscine), the most important of these being hyoscine or 1-scopolamine.

With the development of synthetic drugs, plant products lost their significance. In the last few decades, however, there has been probably more interest in drugs obtained from vegetable sources than at any time in history because of the success with the antibiotics and other plant drugs such as Rauwolfia (for the treatment of mental diseases), podophyllum (a cathartic, as well as for curing cancerous tumors in mice), aloe (a cathartic, as well as for the treatment of atomic radiation burn); veratrum (hypertensive agent), peyote (psychoactive drug) and sundry others. Sapogenins (closely related to steroids), obtained from many members of Dioscoreaceae and Agavaceae, can be converted into cortisone, into male hormone (testosterone) and female hormones (estrogens and progesterone). Sapogenins have many potenial uses in the treatment of theumatoid arthritis and the female hormones are used in contraceptive pills.

Ethnobotany

Ethnobotany deals with studies among the tribals and rural people for recording their unique knowledge about plant wealth and search of new resources of herbal drugs, edible plants and other aspects of plants, including conservation. It has been recognised as a multidisciplinary study comprising may interesting and useful aspects of plant science, history, anthropology, culture and literature.

Importance of Indian Ethnobotany

During the last half - century ethnobotany has more and more been recognised as a valid discipline that can play a material role in the advancement of many aspects of scientific, sociological and historical studies. An increasing number of investigators have been devoting attention to the vast store of knowledge about properties and uses of plants, still existing in native cultures in several parts of the country.

Indian subcontinent represents one of the greatest imporia of ethnobotanical wealth.

Areas in India rich for ethnobotanical study

India with a diverse flora coupled with large number of aboriginal tribes, inhabiting different pockets in the country, offer immense scope of ethnobotanists. Specially the tribal areas in central India, western ghats, and throughout the length and dialects, etc. The northeast India is mainly inhabited by tribals: there are as many as fourteen distinct subtribes in Nagaland state alone. But none of these tribes are however studied in detail.

Recent Studies

Very little organised ethnobotanical work had been done in the country till about 20 years ago. Organised field work and other studies in the subject were started in the Botanical Survey of India. Studies on ethnobotany were undertaken as an official programme in the Economic Botany Section, since its very inception. Dr. E.K. Janaki Ammai initiated researches on ethnobotany in Botanical Survey of India. She studied subsistence food plants of certain tribals of south India and better prospects of Dioscoreas in India. From 1960 Dr. S.K. Jain from Botanical Survery of India started intensive field study among the tribals of Central India. He devised methodology for ethnobotany, particularly to the Indian context.

The publications for this group in the early 1960s triggered the ethnobotanical activity in many other centres, particularly among botanists, anthropologists and medical practitioners, etc. in India, During the last two decades similar work has now been initiated at various institutions such as National Botanical Research Institute (NBRI) at Lucknow. National Bureau of Plant Genetic Resources (NBPGR) at Delhi Central Council for Research in Unani Medicine. Central Council for Research in Ayurveda and Siddha (CCRAS) and some other institutions. A detailed account of ethnobotany of the Tharus in Uttar Pradesh has been brought out by scientists on National Botanical Research Institute. Several Universities Botany departments such as in Mysore, Kashmir, Bubaneshwar, Calicut, Garhwal, Jaipur, Shillong (and many others) have also initiated ethnobotanical work. Another notewrthy aspect is the recognition of ethnobotany as an organised scientific discipline even for the doctoral and post-doctoral work by academic institutions in India.

In 1974 Dr. T.N.Khoshoo, then Director of National Botanical Research Institute, proposed an All India project on Ethnobotany. It took some years to finalise the work programme and the project came into operation from 1982. This project is in operation at NBRI, Lucknow, 4 centres of Botancial Survey of India (Shillong, Howrah, Coimbatore and Port Blair) and some other institutions. Followed by speciall session on ethnobotany in Botanical conferences and the organisation of a Seminar of "Plants in Folklore and Folklife" on the occasion of X International Congress of Anthrpological and Ethnological Sciences, a Society of Ethnobotanists was established in India in 1981 and 1982 respectively.

Several papers presented in these symposia and seminars have been compiled and edited by Dr. S.K. Jain ujderth title "Glimpses of Indian Ethnobotany" (1981). This indicates the increasing interest developed in various institutions in the country for ethnobotanical studies. Apart from several institutions and Universities, many scientists in different Regional circles of Botanical

Survey of India have undertaken ethnobotanical studies in some areas of Andaman & Nicobar Islands, Andhra Pradesh, Arunachal Pradesh, Bihar, Kerala, Madhya Pradesh, Maharashtra, Manipur, Meghalaya, Orissa, Rajasthan, TamilNadu, Uttar Pradesh and West Bengal among the tribes Asurs, Bhils Bhunji, Chenchus, Garos, Gonds, Hoes, Jaintias, Karbis, Khasis, Konds, Khols, Lodhas, Mahlis, Mundas, Maria, Mishmees, Mompas, Mirils, Nagas, Nicobaries, Onge, Oraons, Pharias, Rotha, Saoras, Santals and Shompens, They have so for recorded about two thousand plants used as medicine, food, fodder, fibre, for house building, musical instruments, fuel, oil, seeds, narcotics beverage in material culture and for magicoreligious purposes. Some studies on plant used in veterinary medicines have also been made.

Indian Institute of Ecology & Environment, New Delhi has proposed to make a herbarium of medicinal plants at New Delhi Administration. For authentication of information and future record, voucher specimens are collected for herbarium.

Papers have also been published on certain special aspects such as origin and basis of local names among tribal societies. Studies have also been undertaken in herbaria and musea for locating less known medicinal uses. Some critical studies have also appeared on the information on medicinal plants collected from literature. These studies have brought on record several unkown or little known use of plants. A large number of less known uses, folklore, taboos and myths about plants have also been needed.

Methodologies of ethnobotanical research

Ethnobotany is sometimes eroneously interpreted as merely a thing of folk medical uses of some herbs. But in its totality the subject involves anthropology, sociology, botany and of course, medicinal and economic botany.

Among the chief lines and tools of research in ethnobotany mention may be made of the following.

(1) Field work

Here actual information about plants are collected in the field with the help of reliable local informants.

Some experience is needed to collect unexaggerated and reliable information from the tribals in the field.

Data are collected on the plant parts, used its collection processing, preparation of the drug. Its dosage and administration. For authentication of information and future record voucher specimens are collected for herbarium.

Local names of Plants are recorded. They help in referring to the same plant again in discussions and for future large scale collection activity.

(2) Literature

Another important source of ethnobotanical information is ancient or unnoticed, published or unpublilshed literature. One of the excellent examples of the use of literature in ethnobotanical research is the large compilation of anti-tumor plants from old texts and local folk medicine from all over the world by Hartwell.

With the vast heritage of vedic literature in India, published work can be used as a valuable ethnobotanical resource. Many plants are referred in literature by their common names and it is common knowledge that the identity of a large number of plants referred in ancient literature has been in doubt and several of these are riddles even today. One can recall the case of *soma* of the Hindu epics. About 20 different plant species have been attributed to this name, varying from plants like *Sarcostemma* (a flowering plant) to *Amanita muscaria* (a fungus). Other significant examples of disputed identity are the names *Brahmi, Purnarnava, Jatamansi, Bala* and *Kalpavriksha.*

The main difficulty in fixing the botanical identity of names of drugs in ancient literature arises due to lack of proper description or illustration of the plant. Usually only the local or common names are given, which of course, have not remained same over the years and hence it is difficult to correlate the names of the ancient with the names of the present, though some names are suggestive of the contents of plants.

(3) Herbaria and Musea

Herbarium sheets and field notes on them have also proved to be a very good source of ethnobotanical data. Schultes feels "these reports have several advantages. Unlike much of the literature, they are in great part first hand : they are attached to an actual plant specimen and there can be therefore, no problem concerning the proper identification of the plant; the ethnobotanical data are anchored down, through the information on the specimen label, to a definite locality and often times to specific people who employ the plant."

The most outstanding work of this type is that of Altschul who searched about 25 lakh plant specimens of Harvard University Herbaria. In India, Agarwal and Saha enumerated some sixty five species of medicinal importance from economic Botany Herbarium of the Industrial Section, Indian Museum (BSIS).

Herbarium of the Industrial Section, Indian Museum (BSIS), Jain and Dam have published ethnobotanical accounts from Kanjilal Herbarium (ASSAM).

Tremendous scope exists for ethnobotanical studies from Central National Herbarium (CAL), where specimens from throughout the country are preserved. So far only a few plant groups are scrutinized, and important, valuable informations are brought out.

(4) Archaeological remains

Another aspect of ethnobotany is the study of archaeological remains such as the work of Barlett (1931) and Steward (1976). In India, an attempt was made by Sithole (1970) to describe such palnts from bas reliefs on the gateways of the Great Stupa at Sanchi and the railing of Bharhut Stupa, belonging to the first and second century B.C; respectively.

Currently studies on determining plants found as remains in neolithic period to assess plants used through time and space as food, fodder, and shelter, are carried out in Kashmir University at Srinagar.

Other aspects of Ethnobotany

Ethnobotany does not end in just recording information on plant species used by tribal population, but also deals with aspects of plant conservation.

The practice of Nature Conservation is a very ancient tradition in India. The protection accorded to many plant and animal species such as Peepal and Umber trees is well known. There are also instances of entire biological communities such as all aquatic communities in a rock pool or plant life in a patch of forest receiving protection because of their association with some diety.

Sacred groves are one of the most valuable of such legacies from the primitive practices of nature conservation. These sacred forests are scattered throughout India, which harbour more or less climax type of vegetation and nearly completely immune from human interference on grounds of religious beliefs. Removal of any plant material, even of dead wood from the grove is a taboo. The villagers seem to respect this taboo with complete sincereity even today.

Apart from the preservation of rare and endemic species, the sacred groves also serve the function of preservation of biological diversity of the flora. It is very likely that the trees species in these groves represent genetic variants which are peculiar to that geographical region.

Again, tremendous scope exists on the ethnobotanical studies of the sacred groves scattered on the western ghats and in Meghalaya in northeast India. Some studies have already been carried out in MACS laboratory at Pune on Ethnobotany in relation to nature conservation in Maharashtra and by Botanical Survey of India in Meghalaya.

Ethnobotanical studies also help in revealing the numerous germsplasm stocks in our cultivated plants and vegetables. The tribal pockets living in interior of forests also cultivate numerous vegetable crops and these and are in cultivation from generations and thus represent a distinct genetic stock adopted to local conditions. Some examples of such cultivated crops are *Pipper peeploides* used as a condiment in Khasi hills, Parkia roxburghii grown in country yards for their fruits which are used as vegetables in northeastern states, *Moghania vestita* of Khasi and Jaintia Hills which is mainly grown as a tuber corp, Hodgsonia hetero-clita, a cucurbit which is mostly grown by the Mizo tribals for their oil seeds and edible kernel; *Digitaria cruciata* var, *esculenta,* a minor millet of Khasi Hills and several species of *Alocacia, Colocasia, Amorphophalus* and *Dioscorea* cultivated by several tribes in northeast India.

APPENDIX - I

Abutilon indicum

Efficacious in the treatment of Leucoderma. Infusion of the leaves and roots is used as a cooling agent in fevers. Decoction of the leaves are good as mouth wash and in cases of tender gums. The flowers and leaves are applied to boils and ulcers. The bark and root are good diuretics.

Aconitum ferox

The part used is tuberous root which is antiperiodic, anti-diabitic, anti-phologistic, anti-pyretic, diaphoratic and diuretic. The alkaloid extracted as Aconitine. Even one day's treatment with Aconitine the quantity of sugar in diabetic treatment is reduced. Its ointment is useful in the treatment of rheumatism and neuralgia.

Adhatoda vasica

A good insecticide. Leaves and root expectorant and antispasmodic. The leaves are valuable and antiseptic flowers used in gonorrhoea.

Alangium salvifolum

Fruits are eaten and is used for curing eye diseases.

Alstonia scholaris

The bark is described as tonic, alternative and useful in fever and skin diseases, rheumatism. Juice of the root when taken with milk cures leprosy.

Anogeissus acuminata

Powder of the dry bark of the plant is used orally with the fruit of *Terminalia chebula* and common salt as an appetiser.

Asparagus recemosus

The root is pounded with aerial red root of *smilax prolifera* and is prescribed as a drink to cure urinary disorder as well as discharge of blood in urine. The juice from the fresh root is given orally in dysentery.

Atylosia scarabaeoides
Decoction of leaves is given in dysentery. It also mixed with honey and given to women after childbird.

Azadirachta indica (Neem)
A valuable medicine as an antiseptic, used in the treatment of smallpox.

Small twigs are used as tooth brushes and as prophylactic for mouth and teeth complaints. Extract from leaves are useful for sores, eczema and skin diseases. Boiled and smashed leaves serve as excellent antiseptic. Used as febrifuge.

Baccaurea sapida
Bark is useful in constipation. Adult chew it as such. In case of children about 100 cc of sap is given in early morning.

Balanites roxburghii
The seeds are given in cold and coughs. The bark is used in sores unripe fruit and leaves are pungent, bitter and purgative and are considered to have anthelmintic properties. Oil extracted from the seeds is used in burns and wounds.

Barleria cristata
Two to three leaves are chewed and swallowed for relief from toothache.

Decoction of root with Barleria strigosa and dry fish is given in anaemia.

Bauhinia variegata
The bark is alternative, tonic astingent and is useful in ulcers, skin diseases and diarrhoea. Dried buds are used for cough, bleeding piles, dysentery. The decoction of root is used in dyspepsia and flatulence.

Begonia lacinate
Extracts from stalks are used veneral diseases.

Berberis aristata
The root bark is alternative, tonic stomachic, antiperiodic and alternative. It is diphoratic and antipyretic malaria fever. Used in treatment of eye trouble, eczema and skin diseases. Powered root is used in haemorrhage in piles.

Betula utilis
The decoction of the bark is used as a wash in otorrhoea and pois and wounds and purs. The infusion of barks is used as a carmative. It is prescribed in hysteria.

Cedrela toona
The bark of this tree is a powerful astringent, used in chronic dysentery.

Celastrus paniculata
Seeds are used in rheumatism, paralysis and leprosy. Oil extracted from the seeds is a medicine for beriberi.

Cephaelis ipecacuanha
The root is emetic. It is a specific remedy in dysentery and is used in piles and gonnorthoea. It is also used in cold cough and asthama.

Citus medica
Fruits used in typhoid and they help in digestion. The rind is made into marmalade and is an antiscorbutic. It is made into a preserve and is used for dysentery.

Dicentia thatictrifolia
Extract drom roots is used for stopping amenorhoea.

Dichoroa febrifuga
The root is emetic and febrifuge. The decoction of root is given in fevers. Experiments indicate promising results in

the treatment of Malaria and other fevers without any after effects.

Dillenia indica

The juice of fruit mixed with sugar and water is used as a cooling beverage in fevers and as cough mixture. The bark and leaves are astringent and are used medicinally. The fruit is slightly laxative, but is apt to induce diarrhoea if too freely indulged in.

Drymania cordata

Leaves and plants are smouldered over fire and inhaled through nose in complaints of headache. It is used as a medicine for dogbites, and is also very useful as throat paint hay fever, diptheria and pneumonia.

Eleusine voracana

It is astringent and nutritious. It is stimulant and act as a mild purgative. It is used as a poultice.

Evodia roxburghiana

Root bark boiled in oil are given to improve complexion. The juice of leaves is given in fever. Fruits are stomachic.

Eranthemum platiferum

A paste of the roots of this plant and leaves of Naravelia zeylanica in equal parts by weight is applied on bone fracture. The treatment is renewed each day till the bone heals.

Erythina variegata var orientalis

Warm leaf juice is applied on the chest in case of cough. The bark is used in a medicine applied in eyes, especially for babies. A soup made with bark of this free and leaves of Adhatoda vasica is given to cure cough and cold.

Flacourtia indica

Paste of about 7 gms bark with 2 gms mustard seed is

heated in an earthen pot and given with curd twice daily in dysentery.

Paste of bark is applied on eczema.

Flemingia congesta
The paste prepared from this plant is used for external application ulcers and swellings.

Garcinia Zanthochymus
The seeds yield a concrete oil known as Dampel oil which is a specific remedy in dysentery and mucous diarrhoea. Its soft twigs soaked in water and made into a paste is used for curing boils.

Glycine soja
A decoction of the root is said to possess astringent properties.

Glycosmis pentahylla
The wood is used in snake-bite. Used in treatment of throat trouble. Stem and branches are used as good tooth-brush.

Glynocardia odorata
The oil is in great repute in India as a remedy for leprosy. The seeds are also used for treatment of eczema and other skin diseases; used in cold as clearing agent.

Helicteres isora
Fruits are soaked in boiling mustard oil, this oil is applied extrernally in case of stomach complaints of children. The oil is massaged on the aching body of children.

Hemidesmus indicus
The plant is crushed and the extract is given in fever.

Heteropogon contortus
An oil distilled from the awna: a very small quantity

(equal to the head of a pin) of it is taken with betel leaf to cure asthma.

Either the roots of the plant are chewed by the victim of snakebite or roots are pounded in water and given.

Houttuynia cordata
The Naga people believe that after returning from the jungle they sometimes suffer from muscular pain due to evil spirits. About 200 cc sap of the raw leaf is given three times a day.

Hoya globulos
Leaf ash is repeatedly applied on dogbite till the wound is healed.

Hygorphila auriculata
The leave paste is applied externally in body pain. The leaf extract is given orally in dropsy: it is also used in Jaundice.

Seeds are considered tonic and saphrodisac.

Hypericum patulum
The seeds are aromatic and stimulant.

Hypericum stricta
One or two teaspoonfuls of infusion or dry crushed leaves mixed with hot water and given once a day to expectant mother.

Ipomoea aquatica
Flowers are crushed and the filtrate is put in inflamed eyes.

Ixora acuminata
Underground parts are made into a paste and applied on wounds. Root decoction is a galactagogue: one or two spoonfuls are prescribed twice a day.

Leonotis nepetaefolia
In swelling of the body due to flatulence or, in women, at the time of delivery, a steam bath using a mixture of this plant, Barleria sp. and Hygrophila auriculata is useful.

Leucas aspera
Leaf juice is put into the eyes two or three times daily to relieve burning sensation and redness of eyes.

Root juice is put in the corresponding nostril in Ardhashishi.

Maytenus emarginata
The leaves are used in fermentation of eyes in pain. A leaf decoction is given in cough: three or four leaves are used.

Melastoma malabathricum
The juice of leaves and roots are used in indigestion. The flowers act as a nervous sedative and are used in piles and haemorrhage.

Mentha viridis (Pudina)
Leaves given in fever and bronchits. Decoction used as lotion for aphthae. Dried plants are made into a chutney and is used for treatment of flatulence & good for kidney. It is a flavouring agent and carmative.

Michelia champaca
The bark is used as an abortifacient for two to three months old pregnancy. As an oral contraceptive root (about 6-8 cm long) is made into a paste with 21 black papers and given after menstruation for three days.

Milania cordata
Useful in stomach pain and dysentery: about 150 cc leaf sap is given three times a day.

Millettia caudata
Leaves of this plant and lxora acuminata and roots of Stauranthera grandiflora mixed in equal parts by weight are applied as paste on snakebite.

Micromelum pubescens
The bark of roots, stems and branches is used for treatment of T.B. cases.

Musa velutna
In dysentery, about 100 cc of raw sap of the pseudostem is given three times a day.Also useful for stopping bleeding.

Nyctanthese arbor-tristis
The bark is mixed with that of Terminalia arjuna and is pounded. The resulting paste is rubbed on the body in cases of internal injuries (by falling etc.) it is also believed to knot broken bones.

The Inflorescence and young fruits are pounded in water, and given to relieve cough.

Ocimum basilicum
The ripe seeds of the plant are soaked in water and put into the eyes for bringing out foreign particles.

Olax Acuminata
Leaves are used as a vegetable. It helps in movement of bowels, useful in stomach pain and pneumonia.

Oxalis corniculate
The leaves are stomachic. Its fresh juice is said to relieve intoxication from Datwa and said to be useful in dysentery and prolapsis of the rectum. The juice is also febrifuge, used in treatment of boils. The fresh leaves made into a curry are said to improve the appetite and digestion of dyspeptic patients.

Phyllanthus asperulatus
Leaves mixed with grains of a grass (Panicumsp.) are believed to promote sexual desire.

Phyllanthus emblica
It is one of the best and oldest medicines used in diarrhoea and dysentery. Its juice mixed with honey and milk is given in gonorrhoea and cures whle leucorrhoea in women.

Picrasma javanica
About 10ml. of the bark decoction is given twice a day in dysentery. A large dose acts as febrifuge.

Powder of the dry fruits, about 20 gm twice a day with water, is given in stomach pain.

Piper attenuatum
In common colds, about 25 cc sap of the root is given three times a day.

Plesmonium margaritiferum
The tuber (about 1 cm long) is boiled, coated with boiled rice and swallowed twice daily as a cure for piles.

Plumbago Zeylanica
The roots are washed, and boiled in milk and given to relive muscular pain: they are also tied round the wrist or round the arm for the same purpose.

Plogonum glabrum
Roots of the plant are pounded and given to the victim of snakebite.

Plygonum plebejum
In amoebic dysentery, the plant is crushed and eaten with tips of adventitious roots of Ficus bengalensis and pieces of bark of Butea sp. The whole plant is used in diarrhoea. Eaten as a vegetable, this plant is believed to promote lactation.

Pterocarpus marsupim
The bark is used in dysentery and in red-urine disease: it is also chewed in tongue diseases, or else juice of the bark is applied.

Pterospermum acerifolium
The bark is useful in intestinal complaints, in headache, and in pains. The flowers ar useful in headache.

Rhododendron com-pamlatum
The leaves are useful in colds hemicrania, chronic rheuma-tism, syphilis and sciatica.

Rothia indica
Decoction of the plant (50 gms) with ginger, 3 gm long pepper and 5 gm black pepper and 10 gm leaves of Andrographis paniculata is divided into eight equal parts. It is given twice daily for four days in fever with symptoms of excess bile.

Rus insignis
Vesicant given in colic.

Saccolabium praemorsum
In case bone fracture, past made from the leaves is applied: the leaves may also be tied on the broken part.

Samlalia malabarica
Its gum like juice is used in dysentery. The resin is aphrodisia young pods are used in gonorrhoea.

Sapindus emarginatus
The bark is pounded, a few grammes are soaked in water for two to three hours, the decoction is given twice or thrice to relieve body pain and headache.

If cattle get ulcers infected with worms at the time of delivering calves, the bark decoction is given: the bark is also hung on the animal's neck.

Saprosma ternatum
About 250 ml salted juice of the bark, mixed with black pepper, is given three times a day in indigestion.

About 20 ml. root juice is given twice a day in painful urination.

About 50 ml of the leaf decoction with ginger is given thrice a day in malaria.

Schima wallichi
The bark is a mechanical irritant and vermicide and is used for curing gonorrhoea by the local people.

Scoparia dulcis
The whole plant is crushed and eaten with candy as a digestive.

Leaves are pounded, made into small pills of the size of rice grain and given three each day of weakness of the semen.

A cooling drink is prepared from the seeds by soaking the seeds overnight. Such use of the plant has hitherto been reported among the Halbi tribes of Madhya Pradesh in India, but it is interesting that, in the West Indies, its branches are put into drinking water to make the water cool.

A piece of root is tied to the arm of a nursing woman, which is believed to improve lactation.

Skimmia laureola
Its leaves are smouldered in houses affected with infectious disease like small pox, chicken pox, etc.

Smilax Macrophylla
The root juice is mixed with powdered Birdal (Phaseolus mungo) and made into cakes and given with warm milk for gout and rheumatism. The patient is not allowed to take salt in his lunch.

Two to three roots of the climber, two to three earthworms and three four red crabs are boiled together in mustard oil for about 30-45 minutes. This medicated oil is used twice daily to cure letic wound. Before using the oil, the wound is washed either with the water of green coconut or with water boiled with Azadirachta indica leaves and five roots of Hemidesmus indicus.

Smilaz Prolifera

Children suffering from the disease of wetting at night are cured if they eat their food served on the leaves of this plant, and use the stem as a toothbrush.

Solanum surattense

The juice from crushed fruits is dropped into an aching ear.

Solamum torvum

The root is pounded with tamarind (Tamarindus indica) and given as an antidote to posioning.

The root barks of this plant and that of Achyrathes aspera and the bark of Santalum album are made into pills. The pills are taken three times daily with hot water for malaria.

Tender fruits are fried and eaten with Raghi in nightblindness.

Streblus asper

The bark extract is used in red-urine disease. A bark decoction is taken with warm water in stomach pain.

The latex is applied to the chest in cases of pneumonia: it is also given internally, half a spoonful, mixed with six black peppers, makes enough for two or three doses-one dose being given daily. Latex is also applied in case of reddening of eyes.

Fruits and fleaves are crushed and given as collyrium in eye diseases.

Swertia chirata
It is a favourite remedy in intermittent fevers, acidity and in bilious dyspepsia. It is bitter and tonic and laxative.

Tamarix ericoiaes
A few leaves are cooked with rice and given to children to relieve cough. The decoction of leaves is reported to cure enlarged spleen in children as well as in adults.

Tectona grandis
The oil obtained by distilation of wood chips cures eczema and ringworm.

Tephrosia purpurea
The paste of the root is applied on the belly to cure dyspepsia: it is also chewed to cure colic pain.The juice of the root is mixed with molasses and taken orally to cure stomach pain.

Uraria picta
The fruit is applied to the sore mouths of children.

Viscum album
Resolvent and laxative. It is used by local hill people as pouttice for treatment of muscular pain, due to a fall or injury and in cases of fracture. A decoction of plant is also given in fever accompanied with aching limbs.

Viscum orientale
The juice of the plant is put in the ear to cure pus formation. The ash of the plant is also put in the ear to cure ache.

Woodfordia fruticosa
Flowers are considered a tonic. A paste of the flowers is

given in cough: it cures in two or three days. A decoction of five or six flowers per dose is given with honey for three to four days in nausea and aversion to food during pregnancy. It is also used in small pox.

Xeromphis spinosa
The roots are used for treating gonorrhoea or diseases of urinary tract. The root bark, with the pitch removed, is pounded in water and the paste is given with milk and sugar three times a day.

Xylocarpus granatus
The seed paste is applied on mumps, boils, swollen breast and in toothache.

Xylosma longifolium
The bark is a common household remedy of the Mikirs for stomach pain and intestinal worms. About 5 gm dry powder is used for stomach pain, while a larger dose in early morning is used for intestinal worms.

The Mikirs prepare an extract from the young leaves, which resembles opium and is used with it.

Zanthoxylum budrunga
A decoction of twigs and leaves is given in fever. A piece of bark placed on the tongue promoted salivation.

Zingiber montanum
The rhizome is considered a good tonic and appetiser, it is given with black pepper in cholera and also as vermifuge.

APPENDIX-II

(i) Root/tuberous types
Alocasia macrorrhiza: Boiled corm eaten, also cultivated.

Alpinia Galanga: Rizome boiled in soup like preparations, also cultivated.

Alpinia speciosa: Rizome boiled in soup like preparations, also cultivated (has gingerlike taste).

Asparagus recemosus: Eaten boiled, raw roots in candied form.

Bupleurum falcatum: Corn boiled and eaten, also cultivated Costus specious Starchy rhizomes eaten boiled.

Curcumazedoara: Starchy rhizomes eaten boiled.

Dioscorea bulbifera: Starchy tubers boiled, also of D. hispida, D. publer, D. versicolor, D. oppositifolia, D. pentayphylla and others.

Elecocharis dulcis: Small tubers eaten raw.

Eriosema chinese: Small tubers eaten raw.

Eriosema comjpestrics: Small tubers eaten raw, also boiled.

Houttuynia cordata: Small tubers eaten raw, also in curreis.

Lasia spinosa: Small tubers eaten raw.

Nelumbo nucifera: Rhizome boiled in soups, curries.

Nymphoides indicum: Rhizome boiled in soups, curries.

Oxalis matiana: Small tubers eaten raw.

Polygonatum multiflorum: Starchy tubers cooked in curries/thick soups by Manipur tribals.

*P.Vertivillantum:*Starchy tubers cooked in curries/ thick soups by Manipur tribals.

Polygonum bistorta: Boiled tubers eaten.

Sagittaria Sagittifolia: Boiled Tubers eaten.

*Tacca leontopetaloids:*Boiled tubers eaten, also cultivated.

Vigna vexillata: Tubers of small carrot size, eaten raw and boiled.

A polycephala: Young shoots/leaves acidic, eaten raw, also cooked.

A Solanacea: Young shoots/leaves acidic, eaten raw also cooked.

Asystasia neesiana: Leaves cooked as vegetable.

Bambusa tulda: Young shoots, culms pickled.

Calamus rotang: Young shoots, culms pickled.

Casearia esculenta: Leaves, young shoots eaten cooked by the Khasis, Mikirs, also C. glomerata.

Cirsium lipskyi: Leaves, young shoots eaten cooked by the Khasis, Mikirs and also by Manipur tribals.

Cissus adnata: Leaves cooked as vegetable: also of C. discolor, C. repens.

Clerodendrum colebroo-kianum: Leaves, tender shoots eaten cooked: also of C. indicum.

Codonopsis sp: Leaves, young shoots eaten cooked by Khasis.

Conocepthatus suaveplens: Leaves, tender shoots eaten cooked.

Dendrocalamus hamiltonii: Shoots boiled and eaten.

Embelia gamblei: Young leaves and shoots eaten boiled: also of E. Naguchia, E. subcoriacea.

Enhydra fluctuans: Raw leaves as salad: also cooked as vegetable.

Garcinia lancafolia: Leaves acidic, cooked by Mikirs.

Hedyotis capitellata: Cooked leaves eaten.

Houttuynia cordata: Leaves, young shoots cooked as vegetable, delicious.

*Hygrophila salicitolia:*Leaves and shoots eaten boiled in soups: also of H. spinosa.

Ipomea aquatica: Leaves, shoots cooked as vegetable.

*Lasia spinosa:*Leaves cooked as vegetable, soups and curries.

*Leena indica:*Tender leaves and shoots in curries.

*Lysimachia candida:*Tender leaves and shoots in curries as vegetable by Manipur tribals.

*Maesa chisia:*Young leaves/shoots as vegetable.

Malva verticillata: Young leaves/shoots as vegetable cultivated by Assam tribals.

Medinella rubicunda: Cooked leaves eaten.

Meliosmapinnata: Leaves, tender shoots cooked with fish as vegetable by Mikirs.

Merremia umbellata: Leaves cooked as vegetable.

*Monochora hastata:*Leaves, tender shoots cooked as vegetable.

Mussaenda glabra: Tender leaves eaten raw, as salad/ in chutney.

M. frondosa: Cooked young leaves/sprouts eaten: also of M.roxburghii.

Natsiatum herperticum: Young leaves/shoots cooked as vegetable.

Oenanthe Javanicad: Young leaves/shoots cooked as vegetable: also used raw, flavoury.

Olax acuminata: Young leaves/shoots cooked as vegetable by Mikirs.

Oxalis martiana: Sour leaves eaten raw.

Paedaria foetida: Tender leaves boiled into stews/ curries.

Pavetta subscapitata: Leaves cooked as vegetable.

Phytolacca acinosa: Leaves/shoots cooked as vegetable: also cultivated in higher hills by Arunachal tribals.

Picris hieraciodes: Leaves/shoots cooked as vegetable.

Polygonum bistorta: Leaves/shoots cooked as vegetables, in stews/soups: also other species. Pouzolzia viminea: Leaves cooked by Khasis.

Permna latifolia: Leaves/tender shoots eaten cooked: also of P. obtusifolia.

Teridium aquilnum: Young sprouting fronds cooked as vegetable.

Rhynchotechum ellipticum: Leaves cooked as vegetable.

Rumes maritimus: Leaves tender shoots cooked as vegetable. Soup like preparation.

*Sambucus javanica:*Tender leaves cooked as vegetable soup like preparation.

Sambucus pulcherrima: Tender leaves cooked as vegetable: also S. nigrum, spirale.

Solena heterophylla: Leaves cooked as vegetable.

Solena heterophylla: Leaves cooked as vegetable.

Stellaria: Tender leaves cooked as vegetable.

Typha angustifolia: Young shoots boiled or stewed.

*Vaccinium donianum:*Young leaves cooked as vegetable: of V. serratum, sour and eaten raw.

Zanthoxylum oxyphylla: Tender shoots cooked as vegetable.

(iii) Buds/Flowers/Inflorescence

Alpinia galanga: Flowers eaten raw.

Ardisia griffithii: Flowers cooked by Khasis and eaten.

Bambusa bambos: Young buds cooked as vegetable.

*Bauhinia purpurea:*Flower-buds cooked as vegetable.

*Bombax celiba:*Flower-buds, fleshy calyx, eaten raw.

*Dichopsis polyantha:*Boiled flowers eaten.

*Dillena indica:*Boiled flowers eaten.

*Dioscorea pentaphylla:*Flower-buds cooked as vegetable.

*Ensete superba:*Flower buds cooked as vegetable.

*Eugenia formosa:*Calyx cooked as vegetable.

Indigofera dosua: Flowers cooked as vegetable.

*Monocheria hastata:*Inflorescence cooked as vegetable.

*Nelumbo nucifera:*Flowr-buds cooked as vegetable, also of Nymbphaea spp.

Polygonum runcinatum: Cooked flowers and inflorescence eaten.

Rhododendron arboreum: Flowers (petals) subsweetish, in jams, cold drinks.

(iv) Fruits-ripe and unripe

(a) Ripe fruits

Actinidia calkisa: Acidic pulp eaten.

Aegle marmelos: Aromatic pulp eaten as such or as a drink with sugar.

Aglaiaedulis: Sweet, succulent/fleshy aril eaten.

Alphonsea lutea: Aromatic pulp eaten; also of A. ventricosa.

Ampelocissus barbata: Sweet berries taste like grapes: also of A. latifolia.

Antidesma acuminatum: Sub-sweetish pulp eaten; also of A. khasianu, A. thwaitesiamum.

Aphania ruber: Sub-sweetish pulp eaten.

Apsorisa diocia: Acid fruit aril eaten raw.

Ardisia floribunda: Sub-sweetish berries eaten; also of A. Polycephala. Sweetish pulp eaten delicious.

Celtis australis: Sweetish pulp eaten, delicious.

Citrus hystrix: Acid, juicy fruits eaten raw; also C. medica.

Coriaria neplensis: Insipid fruit pulp eaten.

Cornus capitata: Pulp sweet, eaten raw/made into preserves, also of C. macrophyllo.

Cyathocalyx martabanicus: Aromatic sweetish pulp eaten.

Debregeasia longifolia: Sub-sweetish pulp eaten.

Dillenia indica: Sub-sweetish pulp eaten; also of D. serabrella.

Diospyros kaki: Sweetish pulp eaten; also of D. lancaefolia, D. ramifloro, D. toposoa.

Docynia hookeriana: Fruits sub-sweet, eaten raw, also as preserve; D. indica better in flavour.

Donella roxburghii: Sweet pulp eaten.

Duchesnea indica: Sour pulp eaten, fruit juicy.

Ehretia acuminata: Sweet pulp eaten.

Elaeagnus lotifolia: Subsweetish pulp eaten; also used in tarts and jellies, also of E.pyriformis. E. umbellata.

Elaeocarpus lanceaefolius: Sub-sweetish pulp eaten raw, also of E. Prunifolius, E. varunna.

Embelia sessilifolia: Subsweetish pulp eaten raw.

Emblica officinalis: Fruit eaten raw, also pickled.

Erionotryaan gustifolia: Sub-sweetish pulp eaten.

Erioglossum rubiginosum: Sub-sweetish pulp eaten.

Evodia fraxinifolia: Sub-sweetish pulp eaten, used in chutney.

Feronia limonia: Aromatic pulp sub-sweetish, eaten raw or made into drink.

Ficus auriculata: Insipid fruit eaten raw and preserved; also of racemosa, F. hispada, F. Palmata. F. Semicordata.

Rissistigma polyanthum: A romati, suvacidic pulp eaten: also of F. Verrucosa.

Flacourita jangomas: Pulp with tarty flavour, eaten raw or as preserve.

Froagria nilgherensis: Sub-acidic pulp eaten, fruit very juicy.

Garcinia aftroviridis: Sub-sweetish pulp eaten: also of G. Cowa, G. lanceaefolia, G. xanthochymus.

Gardenia latifolia: Sub-sweetish pulp eaten.

Gaultheria fragrantissima: Sweetish pulp eaten by Khasis.

Gelonium multiflorum: Sweetish pulp eat*en by Khasis.*

Gironniera cuspidata: Sub-sour pulp eaten, also as drink also of G.sclerophylla.

Haematocarpus thompsonil: Sweet pulp eaten, fruit very juicy.

Holboellia latifolia: Sweet pulp eaten, fruity very juicy,

Horsfieldia amygdalina: Fruit aril sweet, eaten.

Hovenia dulcis: Succulent peduncles, inflorescences, taste like pear, eaten.

Leela crispa: Fruit insipid, eaten.

Mahonia acanthifolia: Fruit insipid, eaten also of M.chisia.

Mahonia acanthifolia: Pulp sour, eaten also of M.napaulensis.

Mangifera indica: Pulp sour, sub-sweetish, much relished, eaten raw/pickled, also of M.sylvatica.

Medinella rubicunda: Fruit insipid, eaten.

Melastoma malabathricum: Fruit insipid, eaten, tastes like black berry.

Melodinus monogynus: Berries fleshy, sub-sweetish, eaten.

Mesua ferrea: Fruity insipid, eaten.

Morrus australis: Fruit fleshy, sweet, eaten: also of other Morus spp.

Myica esculenta: Frint eaten raw or made into a drink.

Myrsine capitellata: Fruit insipid, eaten.

Olax nama: Fruit insipid, eaten, also of D. cuspidata.

Parthenocissus himalayana: Pulp sub-sour, eaten.

Pegia nitida: Pulp-sub sour, eaten aromatic.

Phoenix acaulis: Fleshy sweet pulp eaten.

Purnnus cerasiodies Sub-sweetish pulp eaten.

Purnnus cerasiodies Sub-sour/acidic fruits eaten: also of P. cornuta, P.jenkisii, P.napaulensis, P.wallichii.

*Pyrularia edulis:*Fruit tastes like guava, eaten.

Pyrus psahia: Flavour and taste like pear, eaten.

Randia spinosa: Berries eaten occasionally by Assam tribals.

Rhus javanica: (=Brucea javenica): Sub-acidic pulp much relished.

Rosa gigantea: Soup hips eaten.

Rourea minor: Red oril eaten.

Rubus biflorus: Sweet juicy berries eaten, also of S. lasiocarpus.

Salacia roxburghii: Pulpy, gelatinous, eaten.

Saurauja cerea: Sub-sour berries eaten; also of S. napaulensis, S. panduana, S. roxburghii. Solanum brabisetum: Berries pulpy, eaten.

Sorbus aucuparia: Pearlike fruit sub-sweetish, eaten also of S.vestita.

Stixis suabeolens: Insipid fruit eaten.

Syzygium aqueum: Sub-acidic pulp eaten, also S.cumini, S.mappaceum.

Tetrastigma lancelolaria: Sub-acidic berries eaten.

Uvaria Cor ata: Fruit insipid, eaten.

Vaccinium donianum: Sub-sour fruit eaten, also of V, griffithi-antum, V. sperngeii.

Viburnum colebrookianum: Sub-sour fruit eaten, also of V. continifolium V. grandiflorum.

Willughbeia edulis: Acidic pulp eaten.

Ximenia americana: Sub-sweetish pulp eaten.

Zizphus apeatala: Sub-sweetish pulp eaten, also of Z. funiculosa, Z. incurva and others.

(b) Unripe fruits eaten.

Artocarpus heterophyllus: Fruit pickled, cooked as vegetable.

Dillena indica: Fruit cooked in curries.

Elaegnus umbellata: Fruit cooked in curries, also pickled also E. floribundus, E. prunifolious.

Emblica officinalis: Fruit pickled, also made into sweet preserves.

Ensete superba: Fruit cooked in curries.

Garcinia atrovirdis: Dried sour rind in curries; also whole fruit, G. cambogia, G. pedunculata.

Gardenia campanulata: Fruit eaten cooked; also G.tugida, Mangiferasylvatica: Fruity pickled; also cooked in curries like M. indica.

Momordica cochinchenensis: Fruit cooked as vegetable; also of M.dioica.

Morinda umbellata: Fruit used in curries.

-Moringa oleifera: Fruit pickled.

Mucuna prurita: Fruit (Young green pods) cooked as vegetable.

Parkia roxburghii: Fruit (young green pods) cooked as vegetable.

Randia uliginosa: Fruit (young green pods) cooked as vegetable.

Solanum Kurzii: Fruit (young green pods) cooked as vegetable; also of S. spirale and others.

Sterculia indica: Fruit (young green pods) cooked as *vegetable.*

Trichsanthes dioica: Fruit (young green pods) cooked as vegetable.

(v) Seeks

Artocarpus heterophyllus: seeds roasted and eaten.

Bauhinia vahlii: Seeds roasted and eaten.

Castanopsis hystrix: Nuts eaten raw, also of C. indica.

Cirsium lipskhi: Aromatic seeds eaten raw.

Entoda phaseoloides: Large seeds roasted/boiled and eaten.

Euryale ferox: Seeds eaten raw/roasted.

Hodgsonia heteroclit: Seeds roasted and kernel eaten, much liked by Mizo and Manipur tribals.

Mucuna gigantea: Seeds boiled and eaten; also M. Monosperma, M. Prurita.

Parkia roxburghii: roasted seeds eaten.

Sterculia guttata: Roasted seeds eaten; also of S. versicolor and others.

Vigna umbellata: Seeds boiled and eaten; also of V.capensis, V.pilosa.

4 Wildlife Reserves in India

According to the environmentalists there should be at least 33 per cent area of forest in a country. For ecological balance there should be 60 per cent forests on the hilly-mountaineous terrain and 20 per cent forest cover on plains. In India, the data shows we have 22.8 per cent area under forest, but the recent satellite pictures show that we have hardly 10 per cent to 11 per cent forest area. These forests are natural habitat for the wild animals. So the number and variety of wild life is affected by the expansion and shrinking of natural habitat.

India, being the 7th largest country, is the country where all representative animlas are found. There is no single species which is found in neighbourhood, but not in India. There are transition zones, e.g., the fauna of Indian Thar desert is also common in Pakistan. The animals of dry Himalayas (Ladakh) are common in India and Nepal. Other Himalaya and Teria animals are common in India, Nepal, Bhutan. The wild animals of Ganga-Brahmaputra plain and delta are common in India as well as Bangladesh. Similar is the case with peninsular India and Sri Lanka. Indian subcontinent has very important place on the wild life map of the world. There are about five hundred species of mammals, more than two thousand species of birds, many species of fish, reptiles, amphibians and more than 30,000 forms of insects found in this subcontinent. The Asiatic lion, the armoured one horned great Indian rhinoceros, the magnificient Royal Bengal tiger, the massive Indian elephant and the large Indian gaur are the big five of our forests. They are comparable only with their counterpart animals of Africa where heavy Ethiopian lion, big eared Elephant, two horned rhinoceros, etc. are found. India is the only country in the world where lion and tiger both are found. The three species of cats including the spotted, marbled, fishing, desert, golden, jungle, pallas, the caracal types of cats and the lynx are found. The white tiger is very unique to India.

India has twelve species of bats including the flying fox or fruit bat. The flying lemur, which is neither a lemur nor a bat, but bears the same relationship to the shrews as the flying squirrel does to the squirrels. There are four species of apes in the world, among them mighty gorilla are intelligent. Chimpanjee are found in Afirca, orangutan and gibbon are found in Asia. India has the most noisy ape—the hoolock. There are 5 species of monkeys; the rhesus, the bonnest, the lion-tailed, the Assamese and the pig tailed. There are two species of lemures—the slow Loris and the slender Loris found in Sikkim and Bhutan. The long tailed monkeys—Human, the golden langur, the capped and nilgiri langur are peculiar features of peninsula. Hares and Rabbits are widely distributed in subcontinent. They are found from desert to snow capped hills and from open to thick forests. Rabbits are born naked while hares are covered with hair. There are about 1729 species of rodents in the world, they are also widely distributed in South Asian countries. They are squirrels, antelopes rats, rats etc.

Water in Ganges has not flown much when the wild animals were widely distributed in Indian subcontient. Upto Mough period wild flora as well as fauna were abundant. But after the arrival of Britishers, due to expansion of their various economic and commercial activities like mining, plantation, roads-railways construction, etc. and later in on the world wars, a large sum of forests were swept away and the wild animals were killed in very massive amount. In this century many species which were abundant, came in the verge of extinction. So the need of laving some law to protect them was felt. It was also thought that there would be some areas where the animals can refuge themselves in their natural environment. So the concepts of many categories of forests developed such as National Parks, Sanctuary, protected area, reserves etc. The Indian Board of Wild Life, which is responsible for the management, preservation and development of wild fauna and flora, has defined four areas where wild life receives some kind of protection. Under the provision of Wild Life (Protection) Act, 1972, any area can be declared by the State government as a sanctuary or national park in case the concerned area needs protection and conservation for its ecological, faunal, floral, geomorphological and zoological significance. The Central government can establish a sanctuary or national park with the consent of the concerned State governments.

National Park

According to the Indian Board for Wild Life, "a National Park is an area dedicated by statute for all time to conserve the scenery, natural and historical objects, to conserve the wild life there in and to provide for enjoyment of the same in such manner and by such means, that will leave them unimpared for the enjoyment of future generations with such modification as local conditions may demand".

In the most countries of the world, a National Park can only be created by the national, central or federal government of the country. But in India a State government—can constitute a National Park in its State Legislature—preferably subject to certain standards which are now being drawn by the Indian Board of wild life. The history of National parks in India begins in 1936 when the Hailey (now Corbett) national parks of United Provinces (now Uttar Pradesh) was created. The definition of national parks evolved in 1933 at the London conference of African Fauna, according to which a national park is an area placed under public control, the boundaries of which shall not be altered except by competent legislative authority. The area is declared for the protection and preservation for all time of wild animal life and wild vegetation for the benefit and advantage and enjoyment of the general public. In this area hunting of fauna or collection of flora is prohibited except under the direction of park authority.

Sanctuary

The Indian Board for Wild Life has defined a sanctuary as, 'An area where killing, hunting, shooting or capturing of any species of bird or animal is prohibited except by or under the control of highest authority in the department responsible for the management of the sanctuary and whose boundaries and character should be sacrosanct as far as possible'. The Board has further clarified the position by stating that while the management of sanctuaries does not involve suspension or restriction of normal forest operation, it is desirable to aside a completely sacrosanct area within a sanctuary to be known as `Abhyaranya'. It has also indicated that sanctuaries should be made accessible to the public.

In India sanctuary in usually created by an order or gazette notification of State government. So the weakness of a sanctuary is that it can be desanctuarized merely by another order or gazette notification of a State government because it is not safeguarded by any proper legislation. The idea behind a wild life sanctuary and a national park is same, i.e., maximum protection, preservation and conservation of wild animals. But the fundamental difference between the two is that a sanctuary is created by order of a competent authority, who any be the chief conservator of forests or minister of a State, while a national park can be formed, abolished or changed only by the legislation of a State. Therefore, the status and degree of permanency and protection is much higher in a national park than in a sanctuary. In a wild life sanctuary private ownership rights may continue and forestry and other usages also may continue so long as wild life conservation is not adversely affected. However, in a national park all private ownership rights are extinguished and all forestry and other usages are prohibited. Every national park should have the minimum requisites of fauna, flora and scenery etc. There should be sufficient means of management and protection in a sanctuary to be upgraded as national park. Before declaring any forest as a national park, it must be confirmed that the area is of sufficient importance and status to deserve the appellation. Administration of such area must be improved and illegalities such as poaching and unauthorised gazing should be eliminated.

The size of the park must be adequate, the boundaries should be well chosen and buffer zones should be provided against shooting, grazing and cultivation. Both sanctuaries and national parks may be formed out of reserved or protected forests and wherever forest products extraction is unavoidable, there must be an inner sanctuary or *Abhyaranya*, where no disturbances should be allowed. Central government provides assistance to the State governments for sanctuaries as well as for national parks.

A Protected Area

The Indian Board for wild life describes a protected area, 'In many States there may be are as where it may be considered expedient, (i) to afford special protection to wild life, in order to enable species of wild life which are on the verge of extinction to reestablish themselves, (ii) to afford protection to wild life attracted to water impounded in river valley projects and to other irrigation works, (iii) to afford protection to wild life in and around large towns and sacred place.' Such areas may be constituted as protected area by an order of the government who may also fix the degree of protection and period of protection (may be temporary or permanent).

Elaborating this definition Gee has clarified that protection of wild life does not necessarily imply the protection of vermin. Those wild animals and birds etc. which are injurious to other animals, or to the man in short or long term, can be kept under the control or even destroyed altogether in extreme case in certain places. In November 1972 the Indian WIld Life Board chose the tiger as the national animal of India. A national scheme called the 'Project Tiger' was launched to protect the tiger population. At present there are 16 tiger reserves in India, of these 4 are wild life sanctuaries and 12 are national parks. In India there were 63 national parks and 363 wild life sancturries in 1987. 13 Indian subcontinent has grown with British legacy. The formation of many of national parks and sanctuaries were done by them.

Legislation, Development and Impact of Wild Life

After independence Government of India made some plans for the development of forests and their natural habitats. National forest policy was the first step in this direction in which one third of geographic area was aimed to achieve as forest areas (60 per cent in plains and

20 per cent in hill areas) for ecological balance and preservation of wild life. Government of India is conscious for the preservation and development of wild life and has taken many steps in the direction. Among these steps the most important is the financial assistance to States for the development of sanctuaries and national parks.

Plan Allocation for Wild Life Sanctuaries

There is a total outlay of Rs. 360 lakh for the seventh plan period under the centrally sponsored schemes 'Assistance to States for development of sanctuaries' Financial assistance is given to the States after scrutiny of their proposals on 100 per cent basis for items on non-recurring expenditure and on the condition that the wild life sanctuary being financed is under the complete control of that State's wild life wing. The budget provision for the development of 105 sanctuaries in 1987-88 was Rs 200 lakhs, which increased upto Rs. 275 lakh for the year 1988-89.

Plan Allocation for National Park

During sixth plan there was an allocation of only Rs. 99 lakh for the development of national parks, which was raised to Rs. 260 lakhs in Seventh plan period. Under the central scheme of 'assistance to States for development of national parks' given to States after scrutiny of their proposal, on 100 per cent basis for itmes of recurring expenditure and on 50 per cent basis for items of recurring expenditure. The conditions that the national park being financed is under the complete control of the State's wild life wing. The budget provision for the development of 32 national parks in 1987-88 was Rs. 199 lakh and during 1988-89 it is Rs. 192 lakh.

In November 1972 the Indian Wildlife Board chose the tiger as the national animal of India. A national scheme called the 'Project Tiger' was launched to protect the dangerously depleting species. There are 16 tiger projects in India of which four are wild life sanctuaries and 12 are national parks. These tiger reserves get special attention and allocated from centre. In present plan Namdapha, Manas, Palamu, Nagarjun Sagar etc. are important reserves who have got special attention.

The Directorate of Wildlife under the Ministry of Environment and Forest was set up in pursuance of guidelines were given by the National Wild life Action Plan to assist the State government in the proper management of the national parks and sanctuaries. The Directorate is also responsible for assigning the assistance to States. Its effortrs to enlarge the network of wildlife protected areas to adequately represent country's biological diversity were continued during 1988. By the end of 1988 there were 68 national parks and 366 sanctuaries in the country. The Wild Life institute of India has completed a project for preparation of the biogeographic classification suitable for conservation, planning and has submitted a report entitled 'Planning a Wildlife Protected Area Network in India'. The report recommended that at least 4.6 per cent of geographic area of the country should be under national parks/ sanctuaries. At present it is 4.07 per cent.

The Indian Board for Wildlife, the apex statutory advisory body for formulating policies on wild life, the management of various national parks had been evaluated by a committee appointed by the Board and Dachigham national park was adjudged the best national park for the year 1981-'84. The Indian Board for Wild life has played an important role in encouraging and growing consciousness among the people for wild life. It organises various programmes, functions, seminars and workshops for conservation of wild life during wild life week which is celebrated in first week of October.

The British were responsible for the large scale deforestation and mass destruction of the wild life. But they were also the people who became aware of their preservation and tried to regulate various laws. During that period various acts were passed by British administrators. The Indian Forest Act (1879, 1927, 1950) and its adaptations in States was administered by the forest departments and gave basic protection to wild life in reserved and protected forests as follows:

* Prohibition of shooting, fishing and poisoning water, setting of traps and snares; Section 26(i) Reserved forests, Section 32(j) Protected forests.
* Prohibition of killing or capturing of elephants in areas where the Elephant Preservation Act of 1879 is not in force. Section 26(j) Reserved forests, Section 33(j) Protected forests.

Indian addition to the Indian forest act, there was the 1887 act for the preservation of wild birds and game, or the 'Wild Birds and Animals Protection Act' as it became in 1912. This act was applicable to all types of areas and in areas outside the reserves which was administered by the civil and police departments of the States to which it has been extended, though in reality it was a dead letter. The act was applied to few specified wild birds and animals from time to time. Close reasons were declared for them. At that the penalties were fine upto Rs. 50 and for every subsequent conviction, imprisonment upto one month or fine upto Rs. 100 or both. Besides these acts, certain legislations applicable to certain places and time were also in existence, such as the Elephant Preservation Act of 1879, the Indian Fisheries Act 1897, the Assam rhinoceros Preservation Act and the Bengal Rhinoceros Protection Act.

After Independence the Bombay State (Maharashtra) Wild Birds and Animals Protection Act of 1951 was first step and was also recommended as a model for other States. This Act brought all types of areas, whether government forests or waste lands under the same law and the same set of rules, and this undoubtedly makes for simplicity of administration and control. The conservator of forests was empowered as special wild life preservation officer, assisted by Game wardens (Divisional forest officers or Deputy collectors). The main machinery of enforcement is the forest department itself or revenue patel in the absence of forest staff. All forms of shooting and trapping of wild life were licenced. Penalty and imprisonment for offence was also fixed. Many other States also modified their acts in more or less same line.

Wildlife (Protection) Act 1972 is the recent and most scientific protection measures taken to preserve the vanishing species. Under the provision of this act, any area can be declared as a wild life sanctuary or a national park in case the area concerned needs protection and conservation for its ecological, faunal, floral, geomorphological and zoological significance. The Central government can establish a wild life sanctuary or natioanl park with the consent of the concerned State government.

Efforts to control poaching and illegal trade in wild life by strict implementation of this act were stepped up through the four regional offices. Financial assistance to twelve States to the tune of Rs. 30 lakh is being provided to tackle the problem of illegal killing of wild animals, particularly elephant in the southern States and has had a marked impact. To gather timely and correct intelligence about poachers and illegal traders, a system of cash awards both under the central sector and centrally sponsored schemes was introduced. In the cases of offences relating to species included in Schedule I and

Schedule II - Part II of the act, or to hunting in national park or sanctuary, the offender shall be liable for punishment for a term which shall not be less than six months, can be extended to six years and also with fine which shall not be less than five hundred rupees. In case of a subsequent offence the term of imprisonment shall not be less than one year and the fine shall not be less than one thousand rupees. The causes are compundable. Officers not below the rank of Deputy Conservator of forests are empowered to accept a sum upto Rs. 2000/ by way of compensation of such offences. In this act it is also mentioned that the wild animals, meat trophy, uncured trophy, animal article in respect to which an offence has been committed under this act shall be property of the government.

Tourism: Its impact of wild life

Till 1970s no serious attempt was made to exploit tourism potentialities in national parks and sanctuaries. But the enactment of Wildlife (Protection) Act, 1972 was significant step in the encouragement of tourism. Before this law wild life tourism was disorganised and haphazard. The tourists were not enthusiastic to visit the forest knowing that poachers and other elements might have wiped out or driven away the wild animals.

After enforcement of the new measures; the strict restriction on hunting and building of infrastructures through financial aid by central government the torurism began to develop. They were not only encourages for seeing wild animals, but natural flora also. Wild life toruism has made people more conscious of nature conservation. Though it has also been alleged that tourism has led to the degradation of fauna of sanctuaries and parks. But guided wild life tourism can never lead to degradation of wild life sanctuaries and parks.

However, the wild life tourism could not grow substantially in India. In the case of some of national parks in Corbett, Kanha and Palamu. Only in the case of Sariska and Periyer a marginal growth is recorded. Our wild life tourism constitutes hardly 5 to 10 per cent foreign tourists.

There many reasons behind this state, lack of infrastructure like all season roads-hotel accommodations-vehicles and lack of properly trained guides. Public consciousness and enthusiasm about wild life and nature has not grown so much which can boost wild life tourism.

Other parks and sanctuaries which are far from important places, have yet to do a lot for development. So, the impact of tourism on tribal community is also not countrywide but limited to few places.

The decade of 1970s was most favourable for wild animals when these silent creatures were protected under the umbrella or various laws and acts plans and projects. In 1970 tiger hunting was banned when Mr. Kailash Sankla a forest officer and former director of Project Tiger, drew attention to the plight of tigers in International Union for Conservation of Nature and Natural Resources Congress (New Delhi 1969).

The result was the setting up the Project Tiger. The next important step was the Wildlife (Protection) Act of 1972. Thereafter the state-central government tried to pay proper attention on the preservation of wild life. Due to such efforts many species revied for example tiger, black buck crocodile various types of monkeys etc.

Though many cases of effences have also been reported were wild creatures were hunted by poachers for instance since'1982 four tiger and many dozens black bucks were killed in Sariska alone.

There is other face of wild life problem too. A new problem has emerged new, that is the increasing/ increased number of many species of animals have created problems to the people living in nearby villages. This problem was highlighted in the Conference of State forest minister (May'89, New Delhi).

S.N. Medhi, the forest minisiter of Assam was too annoyed because the elephants have killed hundreds of people. They raid the crops and killed people if they come in the way. At present Assam has largest elephant population around 5,000 elephants in 5,000 sq. km.

Other headache is created by crocodiles. In 1975 project crocodiles was launched to save this reptile from extinction. The project was successful and they grew three times. One third of them were released into rivers, but there are still above 7,000 in 32 farms of the country.

The fisherman of those areas where these crocodiles attacked them and absorbed the fish population. In desperation many States proposed that they will sell crocodile skins in lucrative prices abroad. Initially the Environment Ministry agreed for it and even set up a Pilot Project for 'farming Crocodiles' in Tamil Nadu, but the ecologists opposed it and the government was compelled to stop this project.

Similar was the case with other well known projects. The project Tiger succeeded in increasing the population reported that more than 450 people have been killed by tigers since 1978 in Dudhwa (U.P.). Cattle lifting is a common phenomena. The people of Sunderbans Ranthambhor and other reserves are always afraid of these animal terrorists. The Lions, though not much in number are also wandering out of reserve area.

The most unfortunate situation is that the people

who traditionally protected these animals are turning against them. The Bishnoi community around Jodhpur has protected black population for centuries. Now they are in large numbers (around 12,000) and damage their crops.

Similar is the case of nilgai of Rajasthan who became abundant and always raided the crops of villagers. Such incidents have caused the people around sanctuaries and parks to be 'anti-conservationists'. For instance, in Dudhwa, five tigers were poisoned by villagers who were fed up with their cattle being lifted regularly. Similar case of Poisoning the lion in Gir has also come into light, The Director of the Project Tiger Dr. R.L. Singh has admitted that around 40,50 people became victims of tigers annually.* The local people are raising their voices against this project because they are threatened with tigers and feel that land is encroached by tiger reserves where they are deprived and fodder.

Such dilemma has triggered of a nation wide debate about what to do with excessive numbers. Some are advocating for hunting also. But Dr. Ranjit Singh (Additional Secretary Ministry of Environment and Wildlife) a well known conservationist disagrees on this idea becaues if one their killing is allowed it would be very difficult to regulate it. It would lead to misuse and destruction of wild life. Mr. H.S. Pawar, Director Wildlife Institue of Indian Dehradun, has rightly pointed out that local abundance of a species does not mean that it is out of danger. It means that it present habitat is insufficient.

Real problem does not lie in the protection and conservation but in the management and planning of reserves. The original Project Tiger plan envisaged buffer zones around protected core, meant to take care of any pillover of tigers while allowing people from adjoining villages to for age. But so far the project officials have totally failed in preserving the buffer zones.

The reality is that most of India's wild animals are pushed into 363 plus 63 sanctuaries and parks which account for barely 4 per cent of country's geographical area. These forest areas have also been encroached by villagers.

Traditional corridors which enabled the animals to move form one forest to next have been cut off. That has led to very little gene mixing a vital part of natural regeneration. The Gir lions are living in a isolated small patch. They can be wiped out by just one epidemis. Even the protected parks and sanctuaries are not free from human intrusion and cattle grazing.

A study done by Indian Institute of Public Administration found that about 50 per cent of parks and 70 per cent of the sanctuaries, the people are living inside the core zones. Most of buffer zones have disappeared and the fringes of these sanctuaries have turned into major battleground between man and animals.

Now the Ministry of Environment has taken vital steps of indentifying corridors between forests to enable movements of animals.

The Wildlife Institute of India Dehradun has already submitted a blue print for developing corridors in all geographic regions in the country which will provide an additional 42 thousnand sq km protected area to animals. But it seems a difficult task to acquire land from villagers from these corridors and rehavilitating them elsewhere.

The ecologists are also concentrating on the development of buffer zones. In the buffer zones areas for human use would be clearly demarcated. Here with the help of villagers in surroundings areas fuel and fodder would be grown to meet their needs and the remaining parts would be developed for animals to roam around freely.

All these plans raise optimism and one can expect in next coming years Homo Sapiens the wisest creature, will learn to live with these silent creatures. He will not only think about himself but about the irreplaceable nature and nature's inhabitants.

5 Biodiversity

Dense tropical forests, which provide habitat for many little-known species like the ruffed lemur and other ecosystems. Vents, such as mangrove swamp in Costa Rica are still poorly understood by scientists. Such as the sulfide chimneys called "black smokers", have been found to contain at least 16 families of invertebrates that were unknown just five years ago. Recently, an entirely new set of unicellular organisms called picoplankton, with cell diameters between 1 and 4 microns, was discovered. The productivity of marine systems may have been underestimated by 50 per cent because scientists were ignorant of the role played by picoplankton and had no appropriate methods of measuring them.

Another serious problem is that scientists do not understand exactly how the diversity of genes, genotypes, species, and communities influences ecosystem "function". Over the past 100 years, geneticists, taxonomists, evolutionists, and ecologists have accumulated much knowledge about diversity. The information gathered attests to the importance of diversity for the proper functioning of many organisms and ecosystems. However, a comprehensive, rigorous, and general theory of biodiversity is lacking. Because the threat to biological diversity is now great, scientists must learn how living systems are influenced by changes in diversity. Given the rapid pace of landscape transformation worldwide, there is some urgency in obtaining this information. Knowledge of biodiversity is also very important for evaluating the impact of global climatic change.

Fossil records show that drastic environmental change has been a major cause of species extinction. Species losses also derive from mutual interaction, such as competition and predation. Also, because the environment is in a constant state of transformation, some species are always being lost. Some changes in the physical environment are cyclical and repeated, while others are less predictable. (However, even cyclical changes are subject to chance.) In any case, it seems reasonable that *genetic diversity provides organisms and ecosystems with the capacity to recuperate after change* has occurred. The scientific evidence for this hypothesis is not conclusive, however.

Equilibrium in Nature

An ideal state in which every element is in equilibrium, or "the balance of nature," cannot exist. The weather changes constantly; the diversity of plants and animals fluctuates; mountains erode; and lakes get silted in. Yet, this idea of a balance, of an equilibrium in nature, has persisted. Conservation managers, for example, endeavor to reduce disorder and create undisturbed environments. Those scientists who support the notion of a balance of nature maintain that ecosystems, although not in balance now, are *move constantly toward equilibrium*. These scientists argue that ecosystems are prevented from attaining balance because of exterior forces, which are called disturbances. Storms, floods, pests, outbreakes, fires, and human-induced changes are all examples of disturbances that are presumed to keep ecosystems from reaching equilibrium.

The concept of the balance of nature gained credence in the early 18th century when Isaac Newton introduced very successfully the notion that nature could be explained in terms of a few simple laws. According to the Newtonian view, the world is formed by elements that are simple and that respond to regular and *deterministic dynamics*. Today, physicists believe that nature is complex, not simple, and that the Newtonian view is insufficient as a general explanation of how the universe functions. The universe, physicists are finding, is not in balance. *Disturbances and irregularities* of all sorts are no longer seen as aberrations but *as integral parts of nature*. Physicists are also discovering that at the very origin of the cosmos, at the so-called big bang, a singular, irreversible, and complex universe was produced. Similarly, ecologists have been observing and documenting that most, if not all, ecosystems are not in balance.

But why should scientists be concerned with whether an ecosystem is in equilibrium? The answer to this question is crucial for understanding biodiversity. Systems not in equilibrium behave very differently than do systems in equilibrium. Their behaviour can even appear strange and mysterious. For example, when the source of a disturbance is removed from a system that is near equilibrium, that system is expected to return to its previous state. When, in an undisturbed forest, a gap is created by landslide or a falling tree, new seedlings grow in the gap and restore the equilibrium. After a few years, it is almost impossible to tell that there once was a gap. But when the source of change is removed from systems that are not in equilibrium, they do not return to equilibrium. Instead, they adopt a new state. For example, when land used for agriculture in the Amazon forest is abandoned, it grows into a grassland or a savanna and does not return to the original forest. Natural disturbances are a necessary part of the forest ecosystem's function. For example without gaps creating disturbances, most forests cannot renew themselves. And fires, storms, and hurricanes are and have been a part of life since the beginning of the planet; without them, ecosystems could not function properly.

Knowing about the features of non-equilibrium systems is important for proper ecosystem management. Because a non-equilibrium ecosystem does not necessarily revert to its previous state when the convulsion is removed it is often impossible for humans to return ecosystems to their original condition. Therefore, one set of management guidelines will not suffice for all situations. The previous history of changes and the present disturbance regime will determine the consequences of any management.

Origin of Diversity

There are two conditions that cause population diversity. First, new genotypes are constantly cropping up in a population through *mutation, recombination,* and related *genetic phenomena* and through immigration of individuals, their gametes, or their propagules. Second, diversity in the population is eliminated by natural selection and lost through *emigration* of individuals. Every genetic variation, from gene mutations to entire species, will disappear eventually. This loss can be a very fast process, or the variants can survive for a long time. Species that have survived for extended periods include horseshoe crabs, which have been around for 200 million years, and cockroaches, which originated even earlier, in the Carboniferous period. The speed at which new variations originate in relation to the rate at which they are eliminate determines the actual diversity of the system.

In the process of gene mutation, all heritable diversity ultimately arises at the molecular level. Gene mutations are chemical changes that take place in the composition of the DNA (deoxyribonucleic acid) molecule, the chemical substance responsible for heredity DNA is remarkable in that it shows complex chemical behaviour not expected from systems in equilibrium. The most remarkable characteristic of DNA is its ability to regenerate itself, its "autocatalytic" behaviour.

The DNA molecule consists of four kinds of repeating units called uncleotides, which can be arranged in any sequential order. DNA has the curious ability to maintain its physicochemical integrity regardless of the order of the four nucleotides. That is, any DNA molecule behaves like any other of the same length, regardless of the nucleotide content. This characteristic of DNA makes life feasible because if only one arrangement were possible, or if the chemical stability of one arrangement were significantly different from that others, then all DNA molecules would be alike.

The order of the nucleotides uniquely determines the characteristics of the chemical products made by DNA. But the order of the nucleotides in a DNA molecule can change, and these changes are called mutations. The regular appearance of *mutations* gives life its diversity. Mutated molecules of DNA reproduce their changes and make *modified enzymes,* which, in turn, make *altered cells.* Mutated cells result in *modified organisms.*

Thus, new mutations permit the evolution of new characteristics by changing the *structure* and /or *function* of enzymes and other *proteins.* The characteristics of the enzymes and proteins are determined by those of the DNA molecule, but not the reverse — the "information" flows only in one direction. This flow contrasts with that of an ordinary chemical reaction, in which there are forward and backward reactions.

The appearance of mutations is a *random* process. There is no way to anticipate or predict what mutation is going to occur, only that some will. It is natural *selection*, however, that determines whether new mutations get established or are eliminated. The processes of mutation and selection can be looked at in two ways. Mutation and selection might be random, independent processes that are not influenced by the characteristics of the system, in which case no active process of diversity maintenance exists. On the other hand, a certain degree of variety may be required for living systems to function properly. If so, there may be *system feedbacks* that modify the rates of mutation and selection. Below a certain threshold, a system may collapse because of insufficient variety. According to this last view, diversity is actively maintained in a system.

Speciation

Mutation and natural selection generate genetic and morphological variation within a lineage. Yet what is most distinctive in a community is the coexistence of different species, each acting independently and not sharing its DNA or mutations. How, then, does the great variety of species originate?

Speciation is the process that separates genetic variation into distinct units, or species. In speciation, the original population of organisms with similar genes, called a gene pool, is divided into two or more gene pools. Each of these new gene pools then acquires a unique set of characteristics (cellular, tissue, organ, and organismic) through mutation and selection. (The mechanisms that create species have been a focal point of research in the field of organic evolution.) Speciation follows most commonly from the *physical division of a gene pool.* The separation inhibits interbreeding between individuals in the two populations. In time, one or both populations change enough to prohibit interbreeding.

Because the fate of every species is to become extinct eventually, there must be an influx of new species for life to continue. If, presently, there are about 10 million species, and, on average, each species has a life span of 1 million years, then an average of 10 new species must originate each year. On the other hand, if the life span of a species is only 100,000 years (as many biologists argue), then 100 new species must originate each year. Scientists are often aware of a species' extinction, but the appearance of new species is difficult to observe. Because most species are insects and other small invertebrates, most new species fall into these two poorly studied groups. Thus, there are few well-documented cases of speciation.

Evolution and Resource Limitation

Certain individuals in a population survive and reproduce, while others die without leaving offspring. To understand this phenomenon, scientists must first realize that there usually is not enough of some environmental resources (food, water, minerals) to go around. The availability of resources in the environment limits the rate at which individual organisms can reproduce and their ability to survive. Nucleic acid replication qnd cell, tissue, and organism growth and reproduction all require energy and materials, of which there are only limited amounts. Individuals that are efficient at harvesting resources have a higher probability of reproducing and surviving than do those that are inefficient. Mutations in DNA provide species with a constant input of physical and chemical variations, some of which improve the efficiency with which the species selects and harvests resources. New mutations may increase the efficiency of protein synthesis at the molecular level, or they may improve the water uptake by a plant at the organism level. Mutations may even increase the efficiency with which a mutualistic association between a plant and its specific pollinator develops at the community level.

DNA molecules that are dissimilar in the order of their nucleotides usually do not have characteristics that directly increase or decrease their own survival or reproductive capacity. Given the appropriate environment and resources, all DNA molecules reproduce at the same rate despite their nucleotide composition. DNA molecules manifest their distinctive survival rates through their effects on the living organisms in which they are embedded. These organisms, be they single cells living in pure culture or multicellular organisms living in complex communities, will inevitable differ in their ability to garner resources. Therefore, they will vary in their survival and reproductive capacities and, consequently, so will the DNA in their cells.

Natural selection is an optimizing process. It is process that favours efficiency and produces adaptation. It is not a process that involves chance, like the appearance of mutations, but it must operate with the elements that it has. Thus, natural selection never makes a "perfect" organism. Because the environment is constantly changing, a perfect organism, if it could be produced, would soon cease to be perfect as the environment changed. Such systems, in which the optimum state is favoured but cannot be attained or maintained, are called frustrated systems.

Diversity is the result of two opposite actions: the processes that produce new genotypes, new varieties, and new species and the processes that eliminate mutations, variants, and species from the system. Natural selection is primarily responsible for the reduction of biodiversity; it acts through differential reproduction and differential mortality. In order words, the probability of reproducing and the probability of dying are not the same for all organisms. Rather, reproduction and death are related to the organism's characteristics as well as to the environment in which that organism lives.

The Ecological Role Of Biodiversity

How is species diversity linked to ecosystem structure and function, to the different ways in which organisms and population interact with each other, and to those community properties that emerge from these interactions? Communities have structures and properties not possessed by the populations within them that are called emergent properties, which include trophic structure, stability, guild structure, and successional stages.

Many scientists believe that species diversity is essential for the proper functioning of communities and for the emergence of community-level properties. Just as many different DNA-encoded enzymes are needed for a complex organism to function properly, so, scientists believe, are many kinds of species necessary to maintain community structure. But is any diversity sufficient, or are specific mixes of species necessary for the communities and ecosystems of function? This is a very old question in ecology, and two opposing views exist. One view is that a community is formed by the species that happened to arrive first — that the mix of species in a community is a matter of chance: "The vegetation of an area is merely the resultant of two factors, the fluctuating and fortuitous immigration of plants, and an equally fluctuating and variable environment."[11] According to the opposing view, "In any fairly limited area, only a fraction of the forms that could theoretically do so actually from a community at any one time.... The community really is an organized community in that it has a 'limited membership.' "

Are these two very different views really irreconcilable? Not necessarily, if a community is defined as the sum of all the plants and animals that grow together in an area, certain patterns can be observed: Areas with more than 1,000 millimeters of evenly distributed rainfall always contain a woodland; conifers prevail in areas of extremely low winter temperatures; trees in warm climates have broad leaves; and succulent plants are found in dry climates. The same is true for animals. The mix of species encountered in a community is not a random sample of all plants and animals in the world. Yet, under close scrutiny, differences that are difficult to explain can be found between similar localities. Ecologists cannot predict what species will occur in a given climate and soil type. Also, there have been cases where dominant species, such as the American chestnut, have been lost, with apparently little effect in the overall working of the biological community.

The community behaves in a "frustrated" fashion. Two types of processes are at work and neither is dominant. First, there is natural selection. Species in the community are constantly evolving to increase their ability to withstand the rigours of the environment. This process accounts for cold-hardy species near the popes, for grasseating ruminants in savannas, and for fruiteating bats in forests. The individuals that escape their predators and survive other environmental challenges pass on their characteristics to their offspring.

Second, there is chance. When the climate changes, characteristics that were once advantageous can become a burden. Diseases might be introduced, such as Dutch Elm disease, for which local species have no defense.

Hurricanes, tornadoes, floods, and droughts are unpredictable and their frequency changes over time. Chance also plays a role in dispersal. For instances, a species might be absent from a community purely by chance. Thus, both natural selection and chance result in the steady coming and going of species through immigration, emigration, extinction, and gene mutation.

A large number of additional unanswered questions exist regarding the role of biodiversity in communities. For example, can more and more species be packed into a community, or are there upper and lower limits? A related question concerns optimal levels of diversity and the factors that control them. Other issues include the role of different individuals within a population, different populations within a species, and so on.

Do the members of the community collaborate in the efficient use of energy and resources? In a variable environment, the existence of individuals with different characteristics may increase the ways in which the population can respond to change. For example, if all plants of a species had similar water requirements, all of them would suffer water stress in any year that was drier than normal, and such periods would result in significantly reduced seed production. But if there is genetic variation, some individuals might perform above average each year within certain limits of environmental variation. Therefore, seed production would be satisfactory in both wet and dry years. Moreover, genetically adaptable organisms should survive in more variable environments than do genetically uniform populations. Experience with crops shows that highly productive but genetically uniform varieties have more restricted environmental requirements than do less productive but more variable varieties. Also, plantations formed by uniform varieties are more susceptible to pest and disease outbreaks. So, variability within a species seems to be important for long-term survival. Does the same hold for a community, which is formed of species that live together but do not have a common gene pool? Communities with high species diversity may also cope with long-term environmental fluctuates better than do communities with few species. However, the evidence is contradictory. Terrestrial communities in the climatically variable midlatitudes are less diverse than tropical communities in more uniform environments. Also, deep benthic communities are among the most variable communities anywhere though they exist in possibly the most equitable environment on the planet.

Diversity and Niche Structure

According to modern ecological theory, every species in a community occupies a singular ecological niche, or a role in the environment. A niche is thought of as a multidimensional space where each dimension is a characteristic of a species. Because every species has at least one physical or behavioural characteristic that separates it from other species, every species has a unique niche.

In principle, niche theory supplies the theoretical framework to explain the number and types of species in a community. Niche theory predicts that communities that vary in total resources or in the quality of those resources will contain more niches and, therefore, more species. Nevertheless, resources are sometimes partitioned differently in communities with very similar resources. In other words, communities may be have large or small niches, which result in a few "generalist" species (those with board niches) or many "specialist" species (those with small niches). Finally, communities with similar resources and niche sizes may still vary in the number of species because of different degrees of niche overlap. Niche overlap is the degree to which two species exploit the same resource. Communities differ in niche size and overlap for a variety of reasons, such as differences in climatic stability and predictability, spatial heterogeneity, primary productivity, computation predation, and degree of disturbance. Thus, niche theory is unable to predict the patterns of species diversity.

Using niche theory to understand why communities vary in number of species has proven very difficult in practice. Many studies have shown very elegantly how species partition various resources in space and in time, but it has proven very hard to show the precise contribution of each factor. A key question is whether two species can have the same niche. In other words, is there species redundancy in a community? Present theory predicts that two species with identical niches cannot coexist, and empirical studies seem to verify this. Still, the question of how similar two coexistent species can be remains to be answered.

Trophic Diversity

Additional unanswered questions concern the role of diversity in food webs. All organisms get their energy from the sun, either directly or indirectly. At the simplest level, cells of the green tissues of plants and some micro-organisms capture the sun's energy. Organisms also need specific chemicals from the environment to develop. For instance, all plants obtain the materials for development from soil or water and the air. With these materials and energy from the sun, plants create energy-rich,complex chemical compounds, such as carbohydrates, cellulose,starch, proteins, fats, and wood. Animals ingest and extract the energy from these compounds and incorporate some of them into their bodies. Decomposers reduce them to simple ions that eventually are returned to the soil or water. In a cyclic fashion, then, energy and chemical substances move from the environment to plants, to animals, to decomposing microorganisms, and back to the environment. This trophic diversity, or "food web", is essential for the proper operation of ecosystems. Many studies have shown that the trophic structure of ecosystems is complex and varies from one ecosystem to another. Understanding how food webs function is essential for understanding the place of humans in the ecosystem,[19] because humans also derive their energy and materials from plants and animals. Moreover, interpretation of food webs is necessary to understand the effect of toxic compounds released into the environment and the effect of the introduction of species into new areas.

Are diverse ecosystems with complex food webs more stable than simple ecosystems? Stability, in this instance, refers to the ability of an ecosystem to withstand disturbances, including those produced by people. Scientists used to believe that simple systems are less stable than complex ones. Clearly, there is a threshold of diversity below which most ecosystems cannot function. For example, no ecosystem can function without some plants and decomposers. Without plants, solar energy cannot be captured, and without decomposers, substances are not returned to the environment. But, is there a threshold below which present complex ecosystems lose their stability? How many different sorts of plants and animals are needed? and, if there are many of each sort, does the ecosystem become more stable?

Productivity, or the total quantity of energy captured and biomass produced by an ecosystem in a year, is related to the issue of stability. Are ecosystems that are formed by many species more productive than those with few species? It would appear that simple systems, both natural (*Spartina* marshes and high-latitude marine systems) and artificial (agricultural systems), are more productive than diverse ones. However, some scientists fear that these systems are not stable and have argued that society's increasing reliance on the productivity of these simple systems could eventually put humans in jeopardy.

Humans and Species Diversity

Apart from the ethical and aesthetic reasons, is there any reason to fear that human survival is at risk if biodiversity is not preserved? Can humans exist surrounded only by agricultural fields, planted forests, and the like? This question is not easy to answer. In the short term, natural ecosystems can probably lose species without any great impairment of function. For instance, many planted forests are much simpler than the natural forests they replaced. As long as the environment does not change very much, ecosystems can apparently lose many of their rare species without any visible effects. In some cases, common animal species, such as passenger pigeons, have disappeared or been drastically reduced without endangering human survival.

Yet, environments inevitably changes, and sometimes drastically. It is then that the role of obscure species sometimes becomes very important. When dinosaurs became extinct, an obscure group, primitive mammals, suddenly evolved into the dominant role they still play. And now, humans seem on the verge of a drastic changes, one of their own making—the result of global warming induced by increased carbon dioxide in the atmosphere.

Predicate climatic changes pose a series of specific questions relating to biodiversity for which there are presently no answers. What is the impact of elevated atmospheric concentrations of carbon dioxide on tree physiology and forest composition? How does elevated carbon dioxide affect soil organisms and soil nutrients? How is climatic change going to affect insect populations, fire frequency, and wind storm? Coastal zones, in particular, are likely to reflect phenomena occurring over a larger geographical area. Already, the North Sea coast is showing the effects of the industrial and agricultural activity of central Europe, which is hundreds of mile away. The effects of industrialization may be catastrophic for many species, which may become extinct. But a few species, and there is no way of predicting which ones, may thrive in the new environment. Therein lies the scientific importance of biodiversity: It increases the likelihood that at least some species will survive and give rise to new lineages that will replenish the Earth's biodiversity.

Possible Research on Biodiversity

A global survey of diversity is essential to understanding the distribution of life forms on the planet. Biogeographical studies have shown that ecosystem diversity varies greatly from one region to another. Tropical forests, coral reefs, and deep benthic communities are very rich in species: deserts and high-latitude areas are species poor. Many studies have shown that species abundance changes naturally along latitudinal and humidity gradients.

An unprecedented transformation of the natural landscapes of the world is now occurring. Yet, different areas are being transformed at different rates. Some regions, such as the Mediterranean basin, have been altered by people for a long time. Other regions have been influenced only moderately by humans. Also, studies show that the response to disturbance varies greatly from one ecosystem to another. Therefore, it is important to determine how different biomes respond to human disturbance and to document the resulting change in biodiversity and ecosystem function.

Classical biogeography concentrated on the study of species ranges and the causes of their distribution. Given scientists' uncertainty about the number of species and the large number of undescribed species, new and more efficient approaches must be developed to complement past studies. These new approaches should take advantage of remote sensing technology and produce a geographic information system with statistical and analytical capabilities. However, because satellites cannot detect species diversity, new measures that are indicative of diversity and are more easily obtained by remote sensing must be used. For example, satellites can detect standing biomass and productivity, and research should help to establish the relationship between biodiversity and these measures.

Other problems to be studies include the effect that habitat fragmentation has on species ranges and on the probability of extinction and speciation; whether fragmented habitats are more prone to invasion; how invasions influence biodiversity; and the effect of increased geographical isolation on populations and species. Also, research should be conducted to develop a methodology for the systematic comparison of biodiversity across regions. To this end, a biogeographical information system for assessing diversity might use satellite information to identify assemblages of species at different geographical scales for long-term monitoring.

Until now, research in biodiversity has been undertaken by taxonomists at such institutions as the Smithsonian Institution, the Harvard University, Herbaria and Musea, the Missouri Botanical Garden, the Royal

Botanical Gardens in Kew, the Musea d'Histoire Naturelle in Paris, and the Komarov Institute in Leningrad, which are located for the most part in the north temperate zone, while biodiversity is found mostly in tropical countries. Taxonomic institutions in both the north and the south have been in relative decline over the last 50 years. Staffs have not increased or have increased only slightly, and resources allocated to taxonomy have not grown in the same proportion as those devoted to other areas of biology. Consequently, at a time when the conceptual and practical importance of biodiversity is becoming clear, there is a shortage of trained personnel and a lack of necessary funds.

Taxonomists alone cannot answer the questions that have been raised by the massive transformation of the Earth. A collaborative effort by many kinds of biologists, including geneticists, physiologists, and ecologists, is required. As new techniques of genome analysis become routine, even molecular biologists may contribute to identifying genetic variation and its biological role.

Both governmental and non-governmental institutions should collaborate to advance scientific understanding of the significance of biodiversity, and many more resources will have to be made available to fund the collaborative research. The International Union of Biological Sciences, the Scientific Committee on Problems of the Environment, and the Man and the Biosphere Programme of the United Nations Educational, Scientific and Cultural Organization are spearheading research in biodiversity. These organizations are in close contact with the International Geosphere-Biosphere Programme and the United Nations Environmental Programme, to minimize overlap and to coordinate activities. Many national institutions also have active programmes of research in biodiversity, but most of these efforts suffer from inadequate funding. Biodiversity research and conservation figure prominently in the agenda for the 1992 United Nations Conference on Environment and Development in Rio de Janeiro, Brazil, where conference planners are seeking a substantial increase in financial support for research in biodiversity.

It is imperative that biologists and the public learn more about the importance of biodiversity and its role in ecosystem functions. Only an international collaborative effort that is supported by adequate resources can accomplish this task. Increased public awareness of the serious implications of humanity's depletion of biodiversity before its importance is understood is helping to create a climate that may stimulate governments to support national and international efforts in this field.

Strategies Conserve Biodiversity?

Biodiversity is the variety of the world's living species, including their genetic diversity and the communities and ecosystems that they form. Today, this diversity of life is facing serious threats. Scientists concur that, at current rates of tropical forest loss and disturbance, roughly 5 to 10 per cent of all tropical forest species will be lost per decade during the next quarter century.

The extinction crisis is not restricted to tropical forests. Freshwater habitats are being altered dramatically as rivers are impounded and exotic species introduced. In the southeastern United States, for instance, 40 to 50 per cent of the freshwater snails are extinct or near extinction. Oceanic islands, where most of the extinctions of the past several thousand years have occurred, also remain among the most threatened ecosystems on Earth. Some 60 per cent of the plant species native to the Galapagos Islands are threatened with extinction. Although the numbers of species that will be lost are greatest in tropical countries, the proportion of species at risk in the temperate zone is still large. In 1988, the Centre for Plant Conservation in Jamaica Plain, Massachusetts, estimated that more than 10 per cent of U.S. Plant species and subspecies are in danger of extinction in their wild habitats.

The loss of species and genetic diversity and the degradation of habitats and ecosystems represent an enormous cost to this generation and all future generations. Biological resources—the portion of biodiversity that is of actual or potential use to people—provide the basis for most human enterprises. The species being lost today possess unknown food, medical, and industrial uses. The ecosystems being degraded through mismanagement of biological resources are losing their capacity to support the human populations dependent upon them, and this degradation is exacting further costs through soil erosion, salutation of reservoirs, local climate changes, desertification, and loss of productivity. Also, the erosion of genetic diversity among agricultural crops through such practices as the substitution of high-response cultivators for local varieties hinders efforts to breed improved varieties. High response vultivars often handicap small farmers who cannot afford the necessary expensive inputs like fertilizer and pesticides.

The fundamental social, ethical, cultural, and economic values of biological diversity have long been recognized in religion, art, and literature. The great interest of children in nature, the numerous wildlife clubs, the generous donations made to conservation organizations, the political support for green parties, and the popularity of zoos and wildlife films are personal and economic expressions of preference that show that the public does not think of biological resources merely in terms of their cash value. Nevertheless, policies that promote conservation can better complete for the attention of government and commercial decision makers if their advocates demonstrate the economic contribution that biological resources make to development. Even partial valuation of the benefits of conserving biological resources can provide, at least, a lower limit to the full range of benefits that conservation contributes to national economic development.

Three main approaches have been used to determine the value of biological resources:

- Consumptive-use valuation involves assessing the value of resources, such as firewood, fodder, and game meat, that are consumed directly, without passing through a market.
- Productive-use valuation involves assessing the value of products that are commercially harvested and marketed, such as timber, fish, game meat, ivory, and medicinal plants.
- Non consumptive-use valuation involves assessing the direct values of ecosystem functions, such as watershed protection, photosynthesis, climate regulation, and soil production, along with the intangible values of keeping options open for the future and the pleasure of knowing that certain species exist.

Such methods of valuation reveal that biological resources confer much larger economic benefits than is typically appreciated. For example, some 84 per cent of the Canadian population participates in wildlife-related recreational activities (a consumptive use), which provide Canadians with benefits that they declare to be worth $800 million (U.S.) annually. In Sarawak, Malaysia, a detailed field study found that wild pigs harvested by hunters had an annual market value of about $100 million.

Policymakers often appreciate productive-use values more than consumptive - or non consumptive-use values relate to large industries, such as forestry and fishing, which are connected to markets and, therefore, reflected in national income accounts. Yet few people appreciate the magnitude of the contribution of biological resources to national economies. One groundbreaking study that showed how to estimate the dollar value of biological resources and analyzed the contribution of wild plant and animal species to the U.S. economy concluded that some 4.5 per cent of the U.S. gross domestic product is attributable to wild species. In the late 1970s, 4.5 per cent of the U.S. gross domestic product was about $87 billion per year.

Although indirect values of biological resources are the most difficult to estimate, they actually may far outweigh direct values. Species without consumptive - or productive-use values may play important roles in the ecosystem and support species that are valued for their productive or consumptive uses. For example, the retention of a wetlands complex outside Boston, Massachusetts, provides an annual savings of $17 million in flood protection alone —excluding the many other benefits, such as sediment reduction, fish and wildlife production, and aesthetic values, that the wetlands afford area residents. In Thailand, the $1.5-million cost of protecting the watershed of the Nam Pong Reservoir more than justified by the lessened rate of reservoir siltation, which requires expensive dredging.

Already enormous, the value of biological resources to humans will grow as those resources shrink through environmental degradation. Diversity in genes, species, and ecosystems provides the raw materials with which human communities adapt to change. Thus, the loss of each additional species, gene, and ecosystem reduces the ability of nature - and people - to adapt to the changing environment.

Who is to Blame?

Where does the blame for the loss of biodiversity and the degradation of biological resources lie? On the surface, the answer seems clear. The proximate causes include large-scale clearing and burning of forests, excessive harvesting of plants and animals, indiscriminate use of pesticides, draining and filling of wetlands, destructive fishing practice, air pollution, and the conservation of wildlands to agricultural and urban uses.

The root causes of biodiversity loss are found in basic economic, demographic, and political trends. These root causes include the human demand for commodities, such as tropical hardwoods, wildlife, fiber, and agricultural products; the growth in human population, which, even without proportional economic growth, places increasing demands on natural resources and ecosystem processes that are already impoverished and stressed; settlement policies that promote the migration of the growing unemployed labour forces to frontier zones; debt burdens that force governments to encourage the production of commodities that can earn foreign exchange; energy policies that encourage inefficiency in many nations and, in so doing, add to the burden of air pollutants and the risk of substantial global climate change; and inappropriate land tenure arrangements that discourage rural people from investing in substainable agricultural practices.

When biodiversity loss is defined as a result of its immediate causes, the response is to take defensive and, often, confrontational actions, such as enacting laws, closing access to resources, and protecting additional areas. Such responses are necessary in times of rampant over-exploitation, but they are seldom really sufficient to change the underlying social and economic causes. Defining the problem in terms of its root causes can stimulate a more constructive response that seeks cooperative efforts to address the social and economic foundations of resources depletion.

Six main obstacles make it hard to conserve biological diversity:

- National development programmes undervalue biological resources monetarily.
- Over exploitation of biological resources yields great profit for traders and manufacturers (who can externalize environmental costs) while impoverishing the local people who have few other sources of livelihood and who must pay the environmental costs of overexploitation.
- The species and ecosystems upon which human survival depends are still poorly known.
- Scientific research often does not meet the needs of resource and protected-area managers.
- Conservation activities tend to be focused too narrowly.
- Institutions assigned responsibility for conserving biodiversity have lacked sufficient financial and organizational resources to do the job.

These obstacles block progress in both developed and less developed countries. However, because the tropics harbour a major proportion of the planet's biological diversity and because overexploitation of resources in the tropics is linked to resource demand in the industrialized countries, the most pressing needs for action to conserve biodiversity are in developing countries.

Developed countries depend on tropical resources for industrial materials, breeding material, pharmaceuticals, tourism destinations, and numerous other tangible and intangible benefits. However, the exploitation of the tropics by the industrialized countries has yielded great benefits without commensurate investments in conservation and without compensation for the environmental costs of overexploitation. Such factors as cheap labour, artificially low-priced raw materials, inappropriate development aid, and the control of commodity prices and interest rates have encouraged rapid resource depletion and destruction. The situation is worsening continuously through the ramifications of the developing world's debt crisis and related high interest rates.

Steps to be Taken

Conservation will succeed only if both proximate and ultimate causes of environmental degradation and species loss are addressed. The complex threats to biological diversity call for a wide range of responses from many private and public sectors, and the mix of these responses must be adjusted to suit local conditions. Because government policies often are responsible for depleting biological resources, policy changes usually are a necessary first step toward conservation. National policies that deal directly with wildlands management or forestry or that influence resource use indirectly by affecting, for example, land tenure, rural development, family planning, food prices, pesticide use, or fuel consumption can have significant impacts on the conservation of biodiversity. National and subnational conservation strategies can often provide the mechanism for identifying necessary changes.

The best way to protect species is to *protect habitats*, and most national governments have established legal means to do so. These means include creating national parks and other types of reserves (some 4,500 major-reserves exist covering nearly 500 million hectares); enacting local laws to protect particular forests, reefs, and wetlands; restricting private concession agreements; and establishing zoning regulations. The responsibility for managing protected habitats often is spread widely among public and private institutions. Although accomplishments to date are impressive, the amount of habitat protected must be increased by a factor of three if it is to make the necessary contribution to the conservation of biological diversity. Many of these new areas will need more flexible approaches to management than are customarily used in national parks.

Protected habitats will succeed in realizing their conservation objectives only to the extent that the areas themselves are managed effectively and that the land surrounding them is managed in a way compatible with those objectives. Implementing such management typically involves making protected areas parts of larger, regional schemes designed to conserve natural resources, induce sustainable human use of the resource base, ensure that benefits of the habitat protection reach the rural population, and involve the rural population in policy decisions related to the protected areas.

Ex situ preservation programmes, such as zoos, aquaria, seed banks, and botanical gardens, supplement in situ conservation by providing for the long-term storage, analysis, testing, and propagation of a few threatened and rare species of plants and animals and their propagules. Ex situ preservation is particularly important for wild species whose populations are greatly reduced in numbers and serves as a backup to in situ conservation, as a source of material for reintroductions to the wild, and as a major repository of genetic material for future breeding programmes of domestic species. Some ex ritu facilities — notably zoos and botanical gardens — provide important opportunities for public education, and many make important contributions to taxonomy and field research.

Curbing the pollution of the biosphere is the biggest and most expensive conservation effort. Biological diversity is threatened by various forms of chemical pollutions, but ultimately, the gravest threat may be global climate change. Mean world temperatures could increase by about 2^{0}C in the next 40 years with an attendant rise in sea level of between 30 and 50 centimeters. Although the species and ecosystems within protected areas will certainly be affected by climate change, it is unrealistic to expect the boundaries of existing protected areas to be moved to track the changing climate because these areas usually are surrounded by more intensive human land uses. Instead, new forms of intervention will be required to maintain the ecosystems deemed desirable.

Many response to the loss of biological diversity have been supported by international legislation that has fostered global cooperation in conservation, as has the Convention on International Trade in Endangered Species, for example. However, species and ecosystems are still being exploited at rates that far exceed their sustainable yield. Recognizing the growing severity of threats to biological diversity and the increasingly international nature of the actions required to address the threats, the World Conservation Union(IUCN), based In gland, Switzerland, and the United Nations Environment Programme (UNEP), based in Nairobi, Kenya, have begun to prepare an international convention on the conservation of biological diversity. This effort has gained broad support from many governments. For example, in 1988, the U.S. Congress passed a joint resolution, H.J. Res. 648, in support of the conservation of biological diversity.

Any strategy to foster the *sustainable use of biological resources must be based on public involvement*. Indeed, initiatives that do not involve local communities generally are doomed to fail. Indigenous people in many parts of the world are especially reliant on natural resources for their cultural continuity and economic well-being. Their role in conservation should be given particular attention, and they should be given opportunities to participate as major players in the design of conservation programmes affecting their resources. Local people should be closely associated with the authorities responsible for both the management of biological resources and the establishment and management of protected areas.

Information Needs

Effective action must be based on accurate information, and the more widely shared the information, the more likely it is that individuals and institutions will agree on the definition of problems and solutions. Developing and using information is, therefore, an essential part of conservation at all scales, from the local to the global community.

The current state of knowledge about species and ecosystems is woefully inadequate; detailed knowledge is still lacking on the distribution and population sizes of even such large and well-studied animals as African primates. It is self-evident that increasing knowledge about the kind and variety of organisms that inhabit the Earth and the ways that these organisms relate to each other and to humans must be a foundation of conservation action. This basic inventory and research should be carried out simultaneously with field conservation efforts; the two activities can be mutually reinforcing. To gain the necessary knowledge, scientists must

- Conduct taxonomix surveys to document the wealth of the world's plant and animal species;
- Undertake greatly expanded ecological research to determine how the various pieces fit together, explain the population dynamics of species of particular concern, assess the effects of fragmentation of natural habitats, and determine what management steps are required to enable ecosystems to flourish with their full complements of species;
- develop and improve mechanisms for ex situ preservation, including both captive propagation of species and their eventual release into natural or restored ecosystems;
- monitor the changes in species diversity and ecosystem function as human influences become more pervasive;
- assess the ecological differences between relatively large but minimally disturbed ecosystems and ecosystems that have been heavily affected by humans to learn how to enhance productivity and restore degraded ecosystems to a more productive state; and
- conduct social science research to determine how local people manage their resources and how changes in resource availability and land use affect human behaviour.

Government agencies, local communities, and conservation organizations all need information to manage biological resources more effectively! Tools that can help meet this need include basic descriptions of flora and fauna, practical handbooks for field identification, rapid inventory techniques, and refined computer programmes for species and habitat inventory. The need for information about the tropics is particularly important because these areas hold most of the world's biological diversity and are losing species faster than scientists can record them. The highest priority for basic inventory work should be given to the most threatened sites of greatest diversity and local endemism. The information contained in the species in these areas could disappear before humanity even knows what it is losing.

Development agencies should support national efforts to establish local, sectorial, and national information management systems by demonstrating methodologies, providing training opportunities for taxonomists and biologists, and subsidizing the publication of status reports. Universities, research institutions, and nongovernmental organizations need to be strengthened so that they can help governments assess their biological resources. Closer working relationships should be established between museums and other taxonomic-oriented institutions and those concerned with conservation of biological diversity.

The Indian Scene

Is there anything in common between the familiar *neem tree* and a heated global debate over protecting the environment? Or between its overwhelming medicinal and pesticidal attributes and a multi-million dollar industry in a rich and developed nation? Yes, surprisingly. For this hardy tree, taken for granted as part of the Indian landscape, is the focus of scientific research in the USA because of its amazing array of *pesticidal properties*. And as studies exploit the goodness of neem via biotechnology, it is likely that somewhere along the line the indegenous people who discovered the bitter magic of this tree will be forgotten, just as its genes begin to show their might in the dollar avatar. Is this an instance of the sharing of a global resource?

The debate over this biological wealth deepens because most of it, on a global scale, originates in the developing, tropical nations of the Third World, which seem to be at cross-purposes with the developed nations who have adopted a big-brother attitude. The rather verbose draft convention has the laudable objectives of conserving biological diversity by ensuring sustainable resource use, providing funds and transfer of technology-to developing nations for conservation, and the fair sharing of the benefits of research in biotechnology arising from conservation of biodiversity. Lengthy and longwinded, the draft has 40 articles, while the overuse of technical jargon coupled with convoluted sentences makes it almost incomprehensible. In the absence of a complete understanding of all its implications, India, as one of the world's biologically wealthy nations, is treading softly, even as the minister for environment and forests, Mr. Kamal Nath, is voicing his objections.

But what does India stand to gain from the Convention? Will it protect India's biodiversity from the new global order of bio-technological imperialism? India certainly has a lot to protect. "Our country is extremely rich in genetic resources with about 15,000 known species of flowering plants," says Mr.R.S. Rao, Director, National Bureau for Plant Genetic Resources, New Delhi, the country's premier institute devoted to gene conservation and collection.

Imagine, then, the amazing array of candidate genes that must be available for biotechnological manipulation. Says Swarahj who heads the insect research division at the Indian Agricultural Research Institute, New Delhi, "Insects constitute a mind-boggling 80 per cent chunk of all living organisms in India, there being roughly 60,000 different species.

Many of these, like the moth which weaves silk, have revolutionized human civilisation. Then there are the microbes, bacteria, fungi and the multitude of sea life, all vital sources of medicines and life-giving compounds."

The list is endless, chock-full of incredible genes from nature, which is why India wishes to understand fully what she stands to achieve from the Biodiversity Convention.

Technology transfers and financial assistance of course, say the pundits. It is this link between biotechnology and bioresources that has reached some consensus now, says Mr. Kamal Nath in an exclusive interview soon after his return from the update meeting in New York.

It's here that the trouble begins as the patronizing attitude of the West could well mean a 'recolonisation of India'.

Points out Ashish Kothari, an observer from the non-governmental sector at the Convention's last meeting in Nairobi, "Technology should be exchanged and not transferred one way since there are a number of appropriate, indigenous technologies which are already underway in India."

Of course, there's no doubt that India's level of technological development is strong. Even so Mr.V.L. Chopra, Director General, Indian Council for Agricultural Research, New Delhi, says, "It may be wrong to take a blanket view and reject all foreign assistance with regard to technology. Instead there should be a case-by-case assessment of the technologies being transferred with a view to retaining the sound and rejecting the useless. Says Mr. Rana "There are two categories of nations — the 'gene-rich, technology-poor' and the 'gene-poor, technology - rich' nations. India falls into neither of these categories being 'gene-rich and technology-rich.

Why then the complications of this convention? "The very basis of biological wealth," says Mr.Chopra, "has been the general concept that the free flow of germplasm on an international scale should continue." But who will call the shots? Chhatrapati Singh of the Indian Law Institute, New Delhi, an expert on environment law, feels that the convention is yet another example of Western formulation of international agreements which, far from being globally participatory, are mere 'sign-on-the-dotted-line' documents.

Also, the sharing of biological wealth must be judged in totality along with the issues of intellectual property rights, the GATT and patents. Even Mr Kamal Nath believes in the need to protect our biological wealth via IPRs and patents.

Here one comes full circle to the neem tree. India is a treasure-house of living things and has a wealth of genetic resources in its umpteen biographical zones — the deserts, coastal areas, rain forests, teak and sal forests and the meadowlands of the mountains. Does their conservation really require a Convention? Yes, say spokespersons of the ministry of environment and forests, but not a convention which will cost the nation its sovereignty.

There is also the feeling that the Convention will focus more on the utilisation of resources than their actual conservation.

Can we then set our own terms without foregoing our independence? Says Mr.Rana, "One way to check the imperialism being forced on biologically wealthy nations like India is to strengthen our gene banks and research base, inventorise and characterize our genetic wealth and invest it in basic research."

Realising that there are several shortfalls in government programmes, M.S. Swaminathan, President, World Wide Fund for Nature (WWF), launched a people's biodiversity programme. Under the aegis of the Jawaharlal Nehru Community Biodiversity Conservation Movement at the WWF, these projects have managed to involve effectively the local communities in reforestation, nursery work and wildlife conservation programmes all over India. Several such programmes have been highly successful.

Some even suspect the motivation of the unusually generous offers made by certain developed nations with regard to financial technological assistance.

Even as Mr.Kamal Nath declares India's decision to reject the Convention unless suggested changes are made, India might continue to attend the negotiations and pre-Rio meetings. Will India be lured by these attractive offers of monetary help? Or will she continue to guard her national interests and biodiversity from the developed nations' hegemony? Though Mr Kamal Nath protests that India is not on dole, only time will tell how these questions will be answered ultimately.

Meanwhile, the draft convention high-handedly proclaims awards to the originators of traditional methods or using biological wealth.

In reality, there are so many illegal outlets for both indigenous genes and their utilisation methods that losses occur every day. So, if neem is today the catalyst of a dollar machine, it may have been another plant yesterday and might be yet another tomorrow. A faulty conservation strategy with one-sided formulations may only further the damage, making poorer nations the 'beneficiaries' of low-grade, obsolete technologies. These losses may far outweigh the gains from such a Convention.

Yet there are the silver linings. Mr. Kothari feels that with negotiation and substantial modification, the Convention could be a useful tool for the conservation of biodiversity since it would provide a vital link between nations, with proposed global lists avoiding duplication while ensuring the conservation of all species. " The balanced discussions between the 80 participating countries have been a definite step forward in the chequered history of environmental movements," he points out. "Further, we should adopt a holistic view of the Convention without rejecting it outright. After all, one never knows when India might be in dire need of a gene not to be found here."

Though the facts are clear, the implications are not. There are too many spokes in the wheel. One worrying aspect is that a great majority of people — even people in key positions — seems to be ignorant of the convention's text. Its political understones are strong and could prove overwhelming. The fault is perhaps in the preliminary formulation of the draft carried out at the World Resources Institute, Washington, DC. If participatory drafting involving all members — 'contracting parties' as the Convention calls them — had been done, some of the hurdles could have been avoided. And yet, Mr. Kamal Nath is of the opinion that with the G-77 group of developing countries backing India's stand, there is room for optimism.

Meanwhile, the 'fair and equitable sharing of

resources' seems to have become the golden phrase now. So Utopian does this seem that one would imagine it will never materialise. Earlier instances of gene hunting in the tropics by Western nations are far from being rosy stories of North-South cooperation. Perhaps, the goodness of neem needs protection in more ways than one.

6 Natural Resources Conservation in India

In broad terms, a resource, or natural resource, is anything needed by an organism or group of organisms. In other words, a resource is something useful but for humanity what is useful or useless can change because of technology, economics, and the environmental effects of getting and using a resource.

Technology cannot bring back an extinct animal resource or a paved over wilderness area, but it can extend the supply of some resources by improving them, using them more efficiently, or recycling them. For example, a unit of today's steel provides 43% more structural support that it did a decade ago. Today we get 7 times more electrical power from 1 ton of coal than we did in 1900. Similarly the energy needed to produce 1 ton of pig iron has fallen eight fold since 1800. In the United States in 1900 only 10% of the copper was recycled. Today about 40% is recycled. However, while many matter resources such as copper lead and silver, can be recycled, we can never recycle energy resources. Once a fossil fuel resource, such as coal, oil or natural gas, is burned, it is gone forever as a useful energy source. The concentrated energy in the fossil fuel is released as heat, which is eventually dispersed into the earth's atmosphere. From there it flows back into space.

Sometimes technology can solve the problem of a scarce resource by finding a substitute or replacement. For instance as a structural material bronze replaced stone, iron replaced bronze, steel replaced iron and now aluminum and reinforced plastics are replacing steel for some structural uses. As an energy source, animals replaced human muscle power coal (to produce steam) replaced animals and oil replaced coal for many uses. There is a vigorous debate over whether a combination of coal derived energy and nuclear power (obtained from Uranium) or a combination of solar wind and plant (bio mass) energy may soon replace petroleum and natural gas.

In addition to technology resource use is tied to economics. Something is useful as a resource only if it can be made available at a reasonable cost. For example once we deplete the easily available supplies of a resource, we have to look harder and dig deeper to find remaining supplies. In the costs of findings and making a scarce resource available rise, the resource will eventually become too expensive for most people. Higher costs may stimulate a new search for new supplies or make mining and processing lower grade deposits economically feasible. But regardless of what we are willing and able to pay, we can't get a resource out of earth if it is not there.

There can also be an economic limit to recycling. Typically recycling is cheaper than mining virgin materials but only if the material to be recycled is not too widely dispersed. For example, if products made from iron and steel, such as cars and toys are thrown away, buried and widely scattered through use labour and energy costs for finding and collecting the objects may be too expensive to make recycling feasible.

The continueity of a resource may also depend on

the impact its mining, processing, and use has on the environment. Even if affordable supplies of a resource is available, its use (at least for certain purposes) may have to be abandoned if this use seriously threatens human and other forms of life. Sometimes the environmental effects of a resource use can be minimized and cleaned up. But this cost money and in some cases the cost may be so high that we can no longer afford to use the resource.

Natural Resources: Definition

The sum of all physical, chemical, biological and social factors which compose the surroundings of man is referred to as environment and each element of these surroundings constitutes a resource on which man draws in order to develop a better life. Thus, any part of our natural environment—such as land, water, air minerals, forest, rangeland, wildlife, fish or even human population — that man can utilize to promote his welfare may be regarded as a natural resource.

Ramade (1984) defined a resource as a form of energy and/or matter which is essential for the functioning of organisms, populations, and ecosystems. In the particular case of humans, a resource in his words is any form of energy or matter essential for the fulfilment of physiological, socio-economic, and cultural needs, both at the individual level and that of the community.

The functioning of ecosystem or of technological civilization involves a continuous flow and consumption of energy sun being the source of it as regards ecosystem and fuels as regards technological civilization. In the light of this fact, a resource may be defined as one of the various forms of primary energy present in the nature.

The five basic ecological variables — energy, matter,

space, time and diversity are sometimes combinedly called natural resources. As such, laws which govern changes in these quantities have great applicability in the resource use.

Classification of Natural Resources

The Classic subdivisions of resources include renewable and non-renewable resources. Renewable resources are those which are reproducible and are obtained from the biomass of living organism, e.g. forests and fish. Resources which are not reproducible and are obtained from the finite non-living reserves are called non-renewable resources e.g. coal and metals. Solar energy although having a finite life, as a special case, is considered as a renewable resource in an much as solar 'stocks' are inexhaustible on the human scale.

Some authors prefer to classify resources, into biotic (or living) and abiotic (or non-living) resources, forest, agriculture, fish and wildlife falling in the first category and land, water minerals etc. in the second.

A broader classification of resources, as proposed by Owen (1971), is represented in Fig.1

Quantity, mutability and reusability are the attributes on which this classification is based. Inexhaustible and exhaustible are main categories of resources based on their stocks in nature. Inexhaustible resources are further divided into two groups. Immutable and mutable resources, depending upon the possibility of their qualitative degradation as a result of man's activity. On the other hand, exhaustible resources, being limited in occurrence, are vulnerable to both quantitative and qualitative degradation. Thus their availability depends upon the method of use, so these are classed into maintainable and non-maintainable resources. Supply of maintainable resources could be made to last for long through wise use, however, some of them have capacity to reproduce called renewable, while other cannot reproduce-called non-renewable resources. Non-maintainable resources have static supply and when destroyed or consumptively used are irreplaceable, yet some of them have potentiality for reuse-called reusable resources, and rest whose permanent exhaustion certain are called non-reusable resources.

Conservation and management of natural resources

The term "conservation" to many means "hoarding" or to control supply of goods in such a way that some part is left for the nature. Most people interpret conservation as total protection or restriction in consumptive use of resources, therefore, wildlife conservation to them means a total ban on killing of any animal and the forest conservation means a complete moratorium on all feelings the term, in fact, is derived from two Latin words-con (together) and servare (to keep or guard), thus meaning `to keep together'. The dictionary meaning of the word is an act of preservation. The term was proposed by Giffor Pinchot in 1988 (see Smith, 1958), who conceived it from the title of `conservares' given to British officials in India appointed to look after the protection of natural resources.

Prof. H.M. Rose defined conservation as "the optimum allocation of natural, human and cultural resources in the scheme of national development, where by maximum economic and social security will be assured. In 1967 the American president John F. Kennedy looked upon conservation as "the wish use of our natural environment... The prevention of waste and despoilment while preserving, improving and renewing and the quality and usefulness of all our resource" World Conservation Strategy declines conservation as "the management of human use of the biosphere so that it may yield the greatest sustainable benefits to present generation, while maintaining its potential to the needs and aspirations of future generations" thus the conservation is a practice embracing preservation maintenance, sustainable utilization, restoration and enhancement of the natural environment. Summarily, thus the aim of conservation is two fold: (i) preservation of quality of environment, and (ii) to ensure continuous yield of useful material-living or non-living-by establishing a balanced cycle of harvest and renewal.

The concept of resource conservation has now come under a still wider field on environmental management which is, in essence, the process of allocating natural and man made resources so as to make optimum use of the environment in satisfying basic human needs at the minimum and more, as far as possible, for an indefinite future, and at the same time preventing depletion and degradation of resources.

Environmental management being a too vague, general, or trivial subject has not developed as a full-fledged discipline so far, further, in many developing countries it is viewed as being irrelevant and counter productive, while in advanced industrialized countries it is often interpreted as a means of coping with the 'burden' of environmental legislation and problems put forth by 'over-zealous' environmentalists what to talk of under developed world! The global population, however cannot and should not overlook and deny the last deterioration of environmental world over resulting in dwindling natural resources, pollution of air, water and land, extermination of a number of life forms, down going healths status, danger to cultural assets and many socio-economic backlashes. The root of the problem lies in too may people using too many resources wastefully.

Since the total natural resources available are

limited, the way out to the problem seems only through control of population and also the per capital consumption of resources. Because it is difficult to bring down per capital consumption, the recycling and stringent conservation can bring hopes for sustained development without reducing per capital use of resources.

Land

Soil consists of the weathered layer of the earth's crust with living organisms and products of their decay intermingled. Lying on the rocky core of the earth, this complex mixture is one of the four prime requisites for life. Along with sunlight, air and water, soil supports all living beings on land, plants as well as animal. Altogether terrestrial plants have their roots in soil from which they absorb life sustaining water and nutrients. Man and all land animals, in turn, depend on these plants for food directly or indirectly. Foods, vegetable oils and fats, forage for livestock, fiber, wood and various other indispensable agricultural products come from soil. Thus, a nations economic well being is inextricably linked with the fertility and abundance of her soil resource because of mounting demands of swelling population and long mismanagement would put in jeopardy the very survival of man.

Land use and abuse

The land although appears to be an unlimited resource but, in fact, its injudicious use would limit the availability of this indispensable life-support system. Thus, a land use may be poor or good, if in a region of low rainfall grasslands are replaced by wheat crops, a temporary desert will eventually result; or if the human habitation and factories are developed on flood plains, the damage is almost inevitable. These are the instances of poor land use. Conversely, if the grasslands are used for moderate grazing by livestock, and it the flood plains are used for forestry or agriculture, these will imply good land use. Ecological principles have great applicability in land use planning. Classification of use should, therefore, be based on natural ecological features of the land in question viz, soil type, slope, biotic communities etc.

Rural land-use Planning

To a farmer, soil is the essence of survival and his economic well being is firmly linked with the fertility of his land. In rural land-use planning, concentration is chiefly devoted to creating and developing more farmland by removing forests an natural grasslands, drawing marshes and swamps, channeling streams for irrigation, and so on. Unfortunately, no effort is made to save existing prime farmland from degradation by ill-planned development. Rather than exclusive farm, range or forestry oriented planning, the land use should be made under broad consideration of social, economic, and ecological intricacies.

Urban land-use planning

Deterioration quality of urban and suburban environment is to a great extent the result of injudicious land use and is a threat to the whole socio-economic system. Thus planned cities are as necessary as planned farms. Although, the implementation of land-use plan is infinitely more difficult in urban areas than in rural ones because of the human problems involved, yet if not heeded it would lead to chronic pollution, epidemic, and social disorder. In any urban land-use planning it should be seen that: (i) at least a third of all land remains as protective open space in the form of parks, green belts, or refuges etc. including adequate space for efficient treatment of domestic and industrial wastes: (ii) residential housing around villages or cities should be developed in clusters with each unit separated by broad green belts; (iii) typical topographic situation like stream valleys steep slopes and marshes etc. should be retained free from houses and other high-density uses; and (iv) waste disposal areas and industries should be kept apart from human habitations.

Management of Land

From the above account it is clear that major problems emerging out of abuse of land in cities include unsafe and unhygienic dwelling and those in rural areas are of erosion of soil and loss of fertility. While urban problems can be prevented and cured by proper town planning and municipal facilities, the problems relating to rural areas needs long term soil conservation programme.

Prevention and control of soil erosion: The primary purpose of soil conservation is to prevent soil erosion and heal the damage where it has not advanced too far to respond to curative methods. This is possible largely through two means:

1. The land should wear a vegetational cover throughout the year. Plant roots bind the top soil and prevent it from being drifted so easily with water or air. The vegetative cover also slows up runoff water and causes much of it to percolate into ground that promotes more vegetative growth. The vegetation-whether grasses, trees, or cultivated crops-should fit to the capabilities of the soil and water available.
2. Engineering and agronomic practices should be applied conjointly to counteract the erosive force of both water and wind on the soil. These include contour cultivation, cover crops and crop rotation, and contour strip cropping, used alone or in

combination with terracing where it is required. When terraces are properly constructed and adequately supported by approved cropping and tillage practices, they prove to be one of the most effective erosion-control measures applicable to cultivated sloping lands.

Reclamation of eroded lands: Conservationists are divided into two camps over the issue of reclamation of land already arrested by ravines. One camp of conservationists and naturalists believes in the policy of "forgive and forget" and advocates that ravines should be allowed to exist, while the other feels that efforts should be made to regain the lost land through buldozing or some other means. If the ravines are left in there present state, these will spread further and engulf more of productive land. While the violent means to regain the ravines in their near original condition may provoke numerous agencies of nature to take revenge on man. Thus, following considerations are important in negotiating the eroded land:

i. Ravines should be provided with sufficient and suitable vegetative cover.

ii. Instead of agriculture, these lands should be reclaimed for forestry, pasture, or horticulture.

iii. Their deficiency in nutrients and moisture for plant growth should be improved.

iv. Further misuse of such land should be prevented. Over-trampling by man and cattle should be avoided.

v. Past damages should be repaired and recompensed under strict ecological consideration. For instance depressions may be filled up, slopes easend out, cattle, trials fenced etc.

vi. Vegetative cover provided should be protected against reckless destruction by local population.

Water

The origin of water on the earth is not clear so far. However, the current presumption is that the primordial earth had no oceans, and perhaps very little atmosphere. It is believed that the volatile constituents bound in the earth's crust, oozing the surface through volcanos, rock movements, and hot springs, condensed to form the ocean and the atmosphere. This way, perhaps this remarkable combination to hydrogen and oxygen — called water, came into being and eventually became an indispendable component of the earth's environment.

Water moves from the ocean to air to land to ocean in set cyclic pattern called hydrologic cycle. In other words, this cycle represents a continuous process of exchange of water from the earth's surface to the atmosphere and vice versa under the influence of solar heat.

It is estimated that of the total available water 97.3 per cent is contained in the oceans and out of remaining 2.7 per cent most is in solid form. The amount of water actually available in the atmosphere is a very minute fraction and estimated to be only 1×10^{-5} of the world's total water.

The factors which control the process of evaporation and evapotranspiration include temperature, humidity, and wind. It is a continuous process going on day and night from the ocean surface, ground surface inland waters (like lakes and streams), plant and animal surfaces. Rising in the atmosphere this water forms clouds that float around in gaseous or droplet form. The atmosphere above the earth up to varying heights of 10 to 17 km (depending upon latitude and season) is called the troposphere up to which the clouds float and in which all weather changes take place. Water returns to the earth in the form of precipitation, most of which takes place over the ocean and a little on the land surface. Rainfall on the land surface may have following fates: it may percolate through the soil and become a part of ground water regime; accumulate in a pond, lake or reservoir as inland surface water; flow down to the ocean in the form of stream; taken up by plants and animals; or evaporate.

Water Requirements and uses

Goethe rightly said that, "Every thing originated in the water, and everything is sustained by water" Water is needed to fulfill diverse requirements in so many diverse ways. It is vital to life since for all physiological activities of plants and is essential for animals as well. On an average it constitutes 80 percent of protoplasm ranging from 8 per cent in dry dormant seeds to 95 per cent jelly fish. It serves not only to quench the thirst but also to meet the food requirements because it is an essential raw material in the process of photosynthesis through which green plants make food that is used by all trophic levels directly or indirectly. Ambient humidity which is indeed the invisible form of water that surrounds us is necessary to prevent desiccation of terrestrial life forms. Thus, it can be said that water is a biologically essential and it serves as milieu internal as well as milieu internal.

Flood plains of rivers have been the cradle of civilization and centres of population since ancient times. The reason being their high fertility an easy accessibility to water for various uses like drinking, washing and transport. Egypt was developed on the flood plains of the Nile: Babylon on the flood plain of the Tigris: and North Indian on the flood plaims of the Ganges. Even today the Mississippi plain is about the most densely populated area in the USA as flood plains of the Rhine in Europe, and the Yellow river in China.

Besides temperature, water is the other such key

factor that influences the global ecology. Covering about 70 per cent of land surface and circulating in hydrologic cycle water influences weather and climate on any region and thus its flora and fauna.

Agriculture is dependent on water, may it be rainfall, or surface or ground water irrigation. It eventually is a more important commodity in a country like India because her largely agrarian economy depends critically on water. Huge amount of protein rich food in the form of fish shell fish, and prawn is obtained from the inland as well as very potent industry.

The kinetic energy of gravitational flow of water is tapped and transformed into electricity, what is called hydro power. This concept of electricity generation is one of the oldest and cheapest tricks largely used world over. Moreover, it leaves no pollution burden on nature as in case of thermal or nuclear power generation.

Hardly any industry can do without water. It has multifarious application in industrial processes viz, a raw material, solvent, chemical reactant, coolant, and cleaning agent.

"Water, water every where not any drop to drink is the situation not only found amidst sea but also at places where industrial and domestic refuse make the benevolent water sources so severely contaminated with pathogens and or toxins that they are lift impotable. Being the best dilution medium it is frequently used as the cheapest and convenient way for liquid and solid waste disposal. Further is was practically impossible to maintain sanitary conditions without water for it is needed in all common cleaning and washing practices.

Water at many places serves as a very convenient means of navigation. Besides, it adds to the aesthetic value of the landscape and provides opportunities for recreation and sport.

Water being unevenly distributed over the land, influences interstate and international relations and thus is the cause of what is known as hydro-politics.

It can, therefore, be said that while water is essential for life it is significantly linked with social, economic, political and ecological intricacies.

Problems and Management of Water

Management of water implies making the best use of available water resources for human benefit while not only preventing and controlling its depletion and degradation but also developing it in view of the present and future needs. Water, like forest, is a multipurpose resource and it is important to see that its various uses should not conflict with each other and it can be enjoyed in its totality by man and others. Thus, its right allocation, and quantitative and qualitative conservation are the primary tasks before water managers.

Floods, droughts, improper use, pollution and disease transmission are the important problems related to water. Because usable inland water resources constitute a very minute fraction of the world's total water and because man largely depends upon this minute fraction of water for his various activities, it is important to make efficient use ot available water and at the same time prevent abuse of it. This has to be considered not only in giant irrigation projects or industrial and municipal processes but also in every day life. Taps running unpurposefully, leaking water supply pipes, using filtered water for gardens and building construction are a few instances of inefficient and misuse of water in daily life.

Water being the best and convenient dilution medium is the worst affected resource by pollution. To a common man pollution of water means only the drainage of industrial effluents or domestic sewage in natural waters. It may astonish a lay reader that even pure distilled water when released in sea may act as a pollutant since it alters the normal salinity of sea water which adversely affects the marine life. Besides chemical alterations, changes in physical condition of water also have grave effects on the biota of system as during thermal pollution caused by the discharge of heated effluents in a body of water. Pathogens, chiefly entering natural waters through municipal wastes rich in excreta and other organic filth, also render the water unsuitable for human use.

In fact, it is not only the man who has to use the water resources by these are equally important to land animals and more to those fish and other aquatic animals who live in it. Thus pollution of a body of water does prove detrimental not only to the human society but also to regional ecology and the ecology of very aquatic ecosystem. It is, therefore, essential that the water should not be treated as a simple repository of waste disposal, and if inevitable, the wastes only after proper treatment should be released in it under strict ecological consideration.

The qualitative degradation of water is also in a way quantitative depletion of usable resources. Therefore, recycling of waste water after due treatment would relieve the water scarcity to a great extent, especially in regions marred by water shortage.

Future needs and alternate sources of water

The demand for usable water like any other resource, is increasing fast with the continuous increase in world population and continuous increase in per capital demand linked with more of sewage disposal and transport prolific use of water-using gadgets (like air cooling and conditioning devices, washing machines etc.) and increased recreational activities. According to an estimate, the daily per capital demand for water of an American was 600 gallons in 1900 which increased to 1,500 gallons in 1960 and is expected to be increased upto 2,700 gallons by 2000. The fact is no different in developing world. Thus, the question is where will the water needed for fulfilling increasing demands come from? The answer is not so simple and can only be found through long term planning. Possible means of meeting out the in pending water deficit problem include: (i) reclamation of sewage and waste water, (ii) development of ground water sources and surface storages; (iii) long range forecasting of rain; (iv) rain making; (v) transfer of sulfur water, and (vi) desalination of sea water.

Agriculture

The practice through which specific plant species are cared and managed so as to obtain maximum yield of consumable parts of the plant is called agriculture. An agricultural field is, in fact, a cropland ecosystem in which ecological principles involving limiting factors and productivity provide great applicability.

The demand for food production , linked with tremendous population growth, has been increasing fast world over. This has been possible to feed increasing number of mouth through increased agricultural production what is called green revolution. It has resulted chiefly from the industrialization of agriculture, which involves highly domestic plant species, sophisticated chemicals, and large fuel energy subsidies. According to an estimate, to double crop yield through industrialized agriculture, a ten fold increase in fertilizers, pesticides and horse power is required.

The green revolution through industrialized

agriculture (fuel powered) as compared to conventional agriculture (man-and domesticated-animal powered) has following backlashes:

i. Such an agriculture involving huge amounts of fuel, and chemicals in the form of fertilizers, pesticides and weedicides, is the chief cause of environmental pollution.

ii. It may lead to social backlashes when a quick shift from small farms to large industrialized farms takes place. In this process small farmers and agriculture workers are driven off the land to cities where they may not get work, food and housing—a situation of despair and social breakdown.

In retrospect, strong measures should be taken to control the population growth and to improve the conventional agricultural practice with application of current agricultural research under due social and environmental consideration. For a greater and sustained agricultural production following considerations are important:

i. Available land acreage should be properly and judiciously utilized.

ii. Soil fertility should be increased through wise use of fertilizers and organic manures.

iii. High yielding and disease resistant plant varieties should be introduced.

iv. Mixed cropping should be practicised wherever possible.

v. Integrated and balanced use of available water sources, surface and ground water should be made.

vi. Weeds and pests should be efficiently controlled. Integrated pest control practices should be preferred over total reliance on chemical pesticides.

vii. Soil erosion and loss of nutrients should be prevented through maintaining vegetation cover throughout the year.

viii. Exotic varieties should be introduced only after due consideration of long term impacts.

Forestry

A forest is a natural ecosystem having multi species and multi aged trees as dominant community. Forests cover about one-third of the earth's land surface of which about 50 per cent is occupied by tropical forest. Of the total geographical area of India 22.74 per cent is forest as against a minimum of 33 per cent forest cover prescribed under National Forest Policy.

Importance of forests

Functions of a forest may broadly be classified into following three categories:

i. Protective functions: These include the protective role of forests against soil erosion, droughts, floods intense radiation etc.

ii. Productive functions: Forests are the source of wood and many other products like gums, resins, fibers, medicines, katha, honey, bidiwrappers, pulp, paper etc.

iii. Accessory functions: These include the role of forests in recreation, aesthetics, and as habitat of diverge wildlife etc.

Thus, it can be said that forests are important in two ways—ecologically and economically.

Ecological significance: Environmentally, the forests are very important. They help in balancing oxygen and carbon dioxide level in atmosphere, regulate earth's temperature regime and hydrologic cycle. Forests increase local precipitation and water holding capacity of soil, thus preventing drought situation. Vegetation cover provided by forest impedes the velocity of runoff on soil surface checks soil erosion, silting and landslides, thus reducing the danger of flood. The litter derived from fallen leaves maintains fertility of soil by returning the nutrients. Forests also act as refuge of wild animals and provided protection to them against strong, cold or hot and dry winds, solar radiation, rain and enemies.

Economic significance: Wood which has various applications in domestic and industrial processes, is the chief product of forest. Wood when used as fuel, has certain advantages over coal as its sulphur and ash contents are very low but, at the same times, excessive use of fuel wood mens pressure on forests which have many functions to serve. Wood may also be converted into solid (coke) liquid and gaseous fuels. Timber is an important material in building construction and day-to-day uses. Forest provides raw material for various wood-based industries viz. pulp and paper, composite wood, rayon and other man made fibers, sport goods, furniture, boat building matches etc. Miscellaneous products, like bamboos, resins, gums, some oils, fibers, medicines, katha, lack, shellac etc., are also obtained from forests.

Depletion of forests

Increased urbanization, industrialization and mining has entailed indiscriminate felling, of trees has denudation of forests. Nobody will, however, ever learn that who set the first assault on which tree in the history, but this practice of logging down trees for human use has been going on or centuries unabated. Colonists used logs for fire wood and making huts. They also made handles for their tools, furniture and fence posts etc. from it. For them forest had only a present and a past-no future; thus they resorted to ruthless and accelerated exploitation of available forest—a story that still continues with more serious backlashes earth has ever witnessed. India alone is losing more than 1.5 million hectare of forest cover each year and according to Tiwari (1982) 22 million hectare of forests have been destroyed during the last three decades due to over exploitation and misuse and conversion to agricultural fields. The depleted forest wealth would simply deprive the man of economic and environmental values offered by forest, as discussed earlier.

Management of forests

In forestry, the yield is chiefly obtained from the accumulated growth of the past, so this industry has to depend upon the annual growth. Forests are managed to obtain a sustained yield which is obviously possible through sustained growth. Following considerations are important in this direction.

i. A forest should have in correct proportion the trees of different ages and classes so that continuous removal of exploitable produce would not affect the future yield.
ii. Each tree crop should put on normal growth increments.
iii. No obstacle should exist to the continuous harvest of exploitable stems.
iv. Unrestrained felling should be prevented.

For all such planning and decision making, sound ecological knowledge is desirable that include knowledge about climatic conditions, of biotic community, successional trend, life cycle of important tree species methods of arresting growth of undesirable trees and promoting growth of high yield tree species, and biotic inter relationship (such weeds, competitions, pets, parasites and diseases etc.)

Of various tactics employed in forest management,

same deserving special considerations are as follows: (i) restrained felling (ii) block and selective cutting (iii) reforestation (iv) control to disasters and (v) recycling of forest products.

Rangeland

A range is the grassland that provides natural pasture or grazing animals. Grasslands occur where rainfall is too low to support the forest community but is higher than that which results in desert community. Generally, in grassland precipitation ranges between 25 to 75 cm. However, depending upon edaphic factors, such as a high water table, grasslands may also occur in regions of forest climates. Soil of a grassland is characterized by having high humus content.

Over 6, 000 species of grasses, adapted for a great variety of climatic regimes, cover about 20 per cent of earth's surface. In size, they vary from as small plants as about 1 cm. to over 35 cm. tall bamboos.

Importance of Rangeland

Man has been cultivating various grass species as his food and as fodder for his domesticated livestock since time immemorial. History of human civilization the witness of dependence of is man on the seeds of various grasses as his own food, for instance, Indo-Chinese, American and Mediterranean cultures were based on ice, maize and wheat respectively while forage was used as fodder for domesticated livestock.

A range, if not serving as a direct source of food for man, still serves the mankind in a number of indirect ways. Solar energy transformed into food energy by the green cover of the range is available to primary consumers (grazers-domesticated or wild) as food, where in a part of it is converted into meat. Of considerable importance to; man are these herbivores and their products such as meat, milk, skins, hides, horns and wool etc. It can thus be said that at the root of all biological resources that, man utilizes is the forage.

Depletion and Degradation of Rangeland

Man would not destroy an artificial pasture developed by input of his hard labour and money than a natural pasture. Thus, like all natural resources the rangeland, inflicted with human stress is depleted and degraded. Cause of such stress may be ignorance, greed or incompetence of man. Phenomenal growth of cattle industry, on one hand, is responsible for ruthless over grazing. While on the other hand, wide expansion of urbanization, industrialization, and agriculture has resulted in encroachment of rangeland. Often, out of greed too a rancher allows a number of head to graze beyond the carrying capacity of the range being ignorant to the fact that too many sick and weak head would not fetch home as much value as a few head in prime condition of health after grazing on enough good forage.

Community or ecological indicators provide the most practical approach of determining whether or not a range is being utilized properly (Humphery, 1962), Range managers use the term "decreasers" for palatable species vulnerable to grazing their disappearing is an indicator of grazing stress. When such indicators are not heeded unpalatable "weeds". Once established, it may become hard to eradicate the weeds and not only this such vegetation will invite a variety of pest animals like rodents and insects. This makes clear the long-term consequences of overgrazing in a range.

There is increasing need to restore the depleted and degraded rangeland not only out of economic (livestock production) but also ecological consideration since rangeland depletion would also cause problems like soil erosion, spread of pests etc.

Range management

Stoddard & Smith (1955) defined the range management as " the science and art of planning and direction range use as to obtain the maximum livestock production consistent with conversion of range resources."

Efforts to manage land include: regulation of grazing pressure: manipulation of stock (herd) distribution: burning of range: reseeding of rangeland: selection of cattle breeds: and control of weeds, pests and predators.

Wildlife

Even on global level there appears to be a great deal of ambiguity attached to the term "wildlife". Webster's Dictionary defines wildlife as living things that are neither human nor domesticated, specially mammals, birds and fish hunted by man. In American usage, term `wildlife' is used to designate the games and excludes fish, thus they have Fish & Wildlife Services at the federal level. In India too, the common understanding of the people regarding wildlife is of wild animals and shikar, more particularly mammals, birds and reptiles. However, a report entitled Wildlife Conservation in India' (1970) of the India Board for Wildlife states wildlife as the entire native uncultivated flora and fauna of the country. The Wildlife (Protection) Act, 1972, also defines wildlife as any animal, bees, butterflies, crustacea, firsh and moth, and aquatic and land vegetation which form part of any habitat. Thus a comprehensive definition of wildlife may be given as-any or all non-cultivated and non-domesticated life including both, plant and animals.

Importance of Wildlife

The Wildlife resources provide aesthetic, recreational and

economic benefits and at the same time are important ecologically. Nature's millions of forms of life on this earth perform some or the other definite biological function which keeps the delicate balance of our exosphere. In nature flora and fauna go hand-in-hand and one cannot exist without the other. Bird and insects are necessary for the cross pollination of flowers and these with other animals also serve for dispersal and propagation of vegetation. Not only this, the health of vegetation also depends on natural browsing and grazing. Needless to mention that animals, directly or indirectly depend on green plants for their food. Besides food plant also provide shelter to a variety of animal species equally important are the relations among animal populations. If too many herbivores are killed, the predators, depriced of their natural food, will leave forest, become vermin, and prey upon domestic livestock or even man. On the other hand if too may carnivores are killed the herbivores because of release of predation pressure will become too numerous and acquire the status of pest for agricultural crops. Thus most important is the balance of nature of which wildlife in an integral and important part.

Abuse and Depletion of Wildlife

One often comes across the headlines in newspapers like 'Last of the blus whales', save the golden langur Tragedy strikes sea turtles, 'major downfall' in rhino population. Great India bustard on the road of extinction...... and so on which narrate the sad story of abuse and depletion of our wildlife heritage. According to an estimate in last 2,000 years the world has lost, by way of extinction about 160 species of mammals and 88 species of birds.

India is endowed with rich biological heritage and her various seripuries and epies are full of vivid and beautiful account on wildlife. Over 500 mammalian and

1,200 avian species contribute to the riches of the wealth. Unfortunately today the Indian wildlife as that in many other parts of the world, is threatened because of man's unprecedented intervention with nature. Nearly 134 plant species have been declared threatened as against 71,47 and 15 species of mammal, birds, and reptiles respectively (Khoshoo 1984). Great India bustard Choriotis nigriceps confined to the semi-arid regions of Rajasthan and Gujarat, and the white-winded wood dack cairina scutalata inhabiting the Brahamaputra swamps and the flood plains of Assam are both on the birck of extinction.

Wildlife management

The conservation, is the practice embracing preservation, maintenance, substainable utilization, restoration, and enhancement of the natural environment. Wildlife conservation in this sense is the management of nature's living resources if we consider wildlife as consisting both of wild flora and fauna. Since one cannot manage living resources while being indifferent to the non-living environment, the concept of wildlife management becomes the concept of management of natural ecosystems. It also overlaps "forestry" in sofar as it is based on multiple landuse concept and implies management of wild plants as well as wild animals. However, under the present dispensation wildlife management implies management of wild animals as distinct from forestry, and it implies maintenance of habitats suitable to the different species.

In America and Europe wild animals, particularly mammals, are treated as game crops, as do timber in forestry and food crops in agriculture, and hunting is developed as a viable economic activity. In America, wildlife and wildlife related recreation forms multimillion dollar industry. However, in India it is difficult of develop game as in industry because: (i) our forest areas are already overburdened with grazing pressure of domesticated livestock and because of already scarce pasture lands there is little hope of reducing grazing pressure in the forest areas to develop game: and (ii) hunting as a source of recreation is beyond the means of the majority of population.

During past decades the wildlife management implied the understanding, increasing, and controlling of some species. Sanctuaries and National parks were mostly meant of protect a particular species or a group of species and to limited extent for tourism and recreation. Today, however, the emphasis is shifting from the management of species to the management of ecosystem. The objective of wildlife management, as outlined in the World Conservation Strategy, are thus as given below:

i. To maintain essential ecological processes and life support systems in natural ecosystems, on which human survival and development depends. These essential ecological processes and life support systems can be defined to include maintenance of both the biotic and abiotic components of the ecosystems.

ii. To preserve genetic diversity (i.e., the range and variety of genetic material found in the world's wild plants and animals) on which depend the breeding programmes necessary for the protection and improvement of cultivated plants and domesticated animals, as well as much scientific advance, technical innovation, and the security of the many industries that use living resources.

iii. To ensure the sustainable utilization of species and ecosystems. This will include the proper utilization of wild animals for game, forest and grazing, and to achieve proper integration for economy both of rural communities as well as trade and industry. Fortunately, people have been alerted now and the strong measures being taken in favour of restoring, maintaining, and increasing wildlife population given hope that our rich legacy of wild life will not only be saved for future but developed further. Ways and means employed in wildllife conservation include: (i) protection of breeding stock by enacting game laws and developing wildlife refuges; (ii) artificial stocking including introduction of exotics; (iii) control of predators; (iv) habitat improvement, and (v) game farming.

Minerals and Their Characteristics

Minerals are the nonliving substances obtained from the earth through the process of mining. Formation of minerals took place predominantly in weak zones like mountains, folded and faulted regions, lakes, troughs, and continental shelves through geological processes going on since billions of years. They hold some peculiar characteristic, as discussed below, which distinguish them from other natural resources.

(i) Exhaustibility

Minerals are non-renewable or depleting assets and once mined out the deposit get depleted and lost for ever without any chance of replenishment. Refuse or waste, however, can be consumed afterwards through improved benefication technology.

Simply this exhaustible resource cannot be cropped year after year like agricultural products. Therefore, three distinct stages can be identified in the life cycle of a mine: youth-developmental stage; maturity established stage; and old age-declining ore reserves. If the rate of mineral consumption is increasing fast, the knowledge about more reserves and potential substitutability of many minerals is also being improved. As such these predictions may prove outlived.

(ii) Localised occurrence

Minerals are very localised in their occurrence and the total area occupied them is far limited. Most of the mineral deposits are very sporadic in distribution, for instance, Kolar gold fields of India occupy an area of 8 km. by 1.6 km. formations containing coal and petroleum are, in general, bigger as compared to metalliferous deposits. The coal-bearing formation of India, for example, occupy about 4 per cent of the country's total area.

(iii) Unpredictable occurrence

The quantity and quality of a mineral in any region is not easy to predict. Uncertainties prevail with respect to their occurrence, extent of depth, cost of mining (often linked with depth) and the grade.

(iv) Risk involvement

The process of mining involves large number of problems for miners. Abnormal temperature and pressure variations, rock burst, fire and gas hazard, lack of ventilation and drainage, collapse of roof and side walls are the peculiar problems encountered only in mining industry.

(v) Dependence on other factors

Mineral production is largely affected by factors like the changing needs of consumer industries, availability of stock, fluctuation in prices, and reuse of metallic scrap in production process. A technologically poor country possessing richer grades of ores may have to face keen competition in international market with technologically richer countries may, however, possessing poorer grades or ores since in later case the cost of production is considerably low due to improved mining methods.

(vi) International character

Minerals know no political boundaries. Virtually, no country within its own boundaries has adequate production and reserves of all the minerals needed for its industrial development. The unequal distribution of minerals makes the international trade in its inevitable. Some countries are bestowed with monopolistic rights in respect of certain mineral deposits which result in what is know as geopolitics.

Development and conservation of Minerals

Geologists are of general apprehension that the known reserves of minerals may not last for long and most of them will be exhausted within 100 to 200 years. New possibilities of mineral reserves in land and beyond the land (as in the seabed are being explored which may reduce, if not totally remove, the futuristic fears of exhaustion of the resource.

Presently, limenite, rutile, tin ore, diamond, sulphur, coal petroleum, and calcareous and ferruginous sea sands are largely being mined from the sea-coasts and sea beds. In India, ilmenite, rutile and associated zircon, monazite, and sillimanite are obtained from the beach sands of Madras and Kerala coasts. Coral limestone for industrial use like manufacture of calcium carbide, and calcareous sand for cement are being mined from the sea-bed at Tamil Nadu and Dwarka (Gujarat) respectively for quite long time. Indonesia, Thailand, and Burma recover considerable tin ore from the sea- bed, while England recovers coal and Philippines iron from the sea-bed. A number of petroliferous structure occurring off the coast are being mined for petroleum and its products at several places. The ferromanganese nodules or what commonly known as manganese nodules found over the oceanic floors, generally below 4,000 m depth, are perhaps the most spectacular discovery in the field. The most prolific concentrations of such nodules with high metal values have been found in North Pacific between 139 W to 141 W and 4 N. Although precise information about the reserves of manganese nodules is not available but it is estimated that they may run several billion tons and may prove to be potent source of metals like manganese, nickel, copper and cobalt. It is estimated that 1 million tons of manganese nodules from sea-bed provide about 275,000 tons of manganese, 13,000 tons of nickel, 2,5000 tons of cobalt, and 1,000 tons of copper besides other metals like iron, vanadium, molybdenum, silver and lead. Further, it has now been roughly estimated that the metal resources available in molybdenum, silver, and lead. Further, it has now been roughly estimated that the metal resources available in manganese nodules are several times greater than those available in land, such as cobalt 5,000 times, manganese 4,000 times, nickel 1,500 times and copper 150 times.

The conservation of mineral resources does not mean preservation of minerals under ground to be used by future generations but it implies efforts to utilize minerals in best possible way, depending upon the industrial needs and changing technological requirements, so that the life of the deposits is increased without sacrificing out own needs. From the utilization point of view, minerals are placed under two classes: expendable and non-expendable. Expendable minerals have practically no salvage value and ones consumed are lost for ever, e.g., coal, petroleum, gas and most of the non-metallic minerals. Non-expendable minerals are those which have good salvage value and can be used again and again through recycling, except those quantities that are lost due to corrosion and abrasion, e.g., metallic minerals.

Efforts to conserve mineral resources include:

i. Application of efficient methods of mining to take out every possible tonnage lying under ground.

ii. Utilization of unmarketable ores through the innovative methods or ore dressing and metallurgical practices.

iii. Use and reuse of scraps.

iv. Recovering all associated elements as coproduct or by-product.

v. Economic use of minerals.

vi. Substituting the use or rare land costly minerals with those which are abundant and cheaper.

A facility (house, hospital, port) or even in providing a service to society (medical care, public transport). Firewood, fossil fuels (coal, petroleum and its products, natural gas), solar energy, biogas, wind and hydro power, and atomic power are chief energy resources. Of these, firewood and fossil fuel are being used for supply of energy since time immemorial and are known as conventional energy sources. Wind and hydro power, solar energy, biogas, and atomic energy are the new and non-conventional sources of energy. Fossil fuels being limited in supply may be termed as exhaustible sources while others, like solar, hydel, wind and atomic energy, that would be available to man for infinite period are regarded as inexhaustible sources. Firewood and other biofuels (Such as, biogas derived from biological wastes, domestic garbage etc.) are indeed renewable energy sources but exploitation of fire woods has great ecological implications since we are already short of forests and whatever forests available should not be squeezed or thinned. However, use of other biofuels is known to be the best option in the light of two important acts: their steady of even increasing supply at low cost; their use in energy production reduces pollution of environment.

Because of rapid depletion and poor regeneration of the exhaustible sources of energy the society should concentrate on tapping the inexhaustible and renewable sources under strict economic and ecological consideration.

Conventional and exhaustible energy sources

Fire wood: No one can trace back in the history that who are made the first assault that is on which tree. However, it is evident that man has been logging down the trees for various purposes, including to get firewood as an important one, since long. Because, in olden days man's knowledge and access for various natural resources were limited, he ruthlessly exploited his own surroundings to meet out his day-to-day requirements. The story still continues in various parts of the world at slower or faster pace, but often not because of the dearth of knowledge but out of vested interests or because of having no other option. Felling of trees for firewood is still a common practice in interior where it is the cheapest and, at times the only available source of energy. This had led to thinning or denudation of woodland that needs serious consideration from both ecological and economic viewpoints. Obviously, the only wayout to the problem is making available to the villager the cheaper nonconventional energy sources, such as biogas and solar energy in usable form.

Fossilfuel: Forssilfuels (coal, petroleum, natural gas) are the most extensively used sources of energy today. Like for many other natural resources increase in population and per capital energy demand coupled with the industrialization at a faster pace, are the factors responsible for depletion of fossil fuel world over. According to a report the earth's original stocks of coal and petroleum were about 6,000 billion tons and 1,250 billion barrels respectively, of which by the year 1960,100 billion tons of coal was mined and 100 billion barrels petroleum extracted. The story of natural gas is no different which has acquired the status of irreplaceable commodity of domestic use in modern times. Petroleum and natural gas today fulfil 60 per cent of the world's total energy requirements. It is evident that if the exploitation and consumption of these resources continues at the present incredible rate, there supply may last only for a few more decades.

Non-conventional and inexhaustible energy sources

The rapid depletion conventional energy sources has prompted governments and people to concentrate on finding and tapping some non-conventional energy sources that may last for long. In fact, inexhaustible energy sources, like solar, hydel, tidal, wind and atomic power, can only bring hope for the sustained socio-economic development of humanity for years to come when exhaustible sources will become either inaccessible or too costly to afford.

Solar energy: Research and application in the field of solar energy have opened new vistas in the direction of fulfilling world's future energy requirements. It has especially drawn attention for its practically free steady supply and pollution free use. This energy may be transformed into heat using reflectors and condensers, or into electric energy using solar photovoltaics. The resource can especially be tapped more effectively in the regions where there are long sun-shine hours.

Solar heaters and cookers are becoming popular these days in industry and household. Passive solar architectures provide comfortable housing in extreme cold and hot climates. This engineering has also been successfully applied in warehouse and cold storages. A multipurpose solar dryer-cum-warehouse is under construction at Ganaur (Haryana, India) where chillies, potato-chips and other foodstuffs could be dried and stored. Solar energy has also been effectively used in desalination of water. College of Engineering Pune has designed a solar furnace that can generate heat upto 2,000⁰Cu which can be send in various metallurgical and other purposes.

Department of Non-conventional Energy Sources of Govt. of India has prepared a perspective plan that envisages generation of energy through non-conventional sources, chiefly solar to the extent of 250 million tons of coal replacement per annum including 15,000 MW of electricity by the year 2001. Indian's first solar power station, soon to be operated near New Delhi, would be a great breakthrough in this direction. Rural electrification through solar photovoltaics is already in operation at some places in the country. Moreover, photovoltaic irrigation pump sets have also been installed at some places.

Wind power: The power of wind since olden days is beings used in running mills, irrigation of fields, and carrying out other farm activities, but its real potential is not so far being realized. According to inventories of Department of Non-conventional Energy Source of Government of India 20,000 MW electricity can be generated from the wind alone in the country. Wind power plants are set up in various parts of the country but they still account for the total capacity of 3.3 MW only. Obviously, this potent resource is still to be tapped optimally.

Hydro power: Use of gravitational flow of water in generation of electricity, what commonly known as hydro power or hydel power is quite common in various parts of the world. India's, first hydro power station, having a potential of 300 KW, was constructed in 1897 at Darjeeling. Later on, many stations with far high potentiality were set up. The country's economically exploitable hydro power resources have been estimated at 41 million KW (or 41,000 MW) of which only 18 per cent of the potential was utilized by 1979. The Brahmaputra basin has the highest hydro power potential and nearly 30 per cent of he country's total; next to it rank the Indus, Godavari, and Ganga basins respectively.

In world, a total $4{,}910X10^9$KW electric power is generated of which hydropower forms of fourth. In highly industrialized countries like USA, USSR, Canada, Japan, Furnace, Sweden, and Norway hydro power is nearly fully developed whereas in the rest of the world it is yet to be developed fully.

Hydel power is considered to be the cheapest source of electricity with the cost of generation ranging from 2 to 3 paise per unit, compared to a minimum of 5 paise per unit of electricity generated by coal based thermal power plants. Although the initial cost of hydel project is generally 25 per cent higher but owing to lower recurring cost, the cost of generation of hydropower is the lowest. A part from such economic considerations, development of such projects is beneficial for irrigation and other purposes too.

Tidal power of water in costal areas of India, however not so far being realized. According to an estimate a tidal power potential of 9,000 MW has been identified of which 7,000 to 8,000 MW is in the Gulf of Cambay, 800 to 1,000 MW in the Gulf of Kutch and rest in sunderbans.

Atomic power (Nuclear power): As already delineated, the fossil fuels beings exhaustible, such sources of energy are not to last for long with the current trend of consumption. Atomic power appears to be the only hope for large scale energy requirements when fossil fuels are exhausted. The reserves of atomic fuels are said to be over 10 times greater than fossil fuels and the advantage of former is that their very small amounts would produce enormous amount of energy. A ton of Uranium ^{235} for instance, would provide as much energy as 3 million tons of coal or 12 million barrels of oil, Atomic energy has its application not only in generation of electricity but has successfully been used in marine vessels, spaceships and even in chemical and food processing industry.

Important constraints in atomic energy generation are: (i) cost of construction and maintenance of plants; and (ii) disposal of radioactive wastes cost of construction of an atomic power plant and its maintenance is very high as compared to any other electric power plant. For running one atomic power plant unit of 235 MW capacity, heavy water of a worth Rs. 10 core is required per year. Further, as a consequence of frequent failures, the running cost of such plants come to be very high. Moreover, after the working capacity of an atomic power plant is over (in about 25 years), these have to be destroyed so as to avoid radiation spillage. Surprisingly the cost and time for destruction of these plants is almost equal to that for construction. Disposal of radioactive wastes is such a problem that still has no satisfactory solution for long term.

Bioenergy (Energy from biological matter): Organic wastes provide an important renewable source of energy. Biogas generated from the decomposition of organic wastes, like manure and domestic wastes is becoming increasingly popular as a source of heat and light in many parts. It is considered to be advantageous in view of its relatively cheaper supply, and use of organic wastes in its generation reduces the otherwise impending threat of pollution due to their release in environment. As such, it serves two purposes; fuel production and waste disposal. It has more of practical applicability in villages where organic waste in the form of cattle dung, is available in plenty. Biogas so generated is thus economic and convenient to use as compared to conventional practice of burning of dry dung-cakes. Further, the dung-cakes serve only as a source of heat for cooking and allied and purposes while biogas has multifarious applications. The slurry so produces in a biogas plant can well be used as manure for agricultural fields.

Under the National Biogas Development Programme 8.2 lac biogas plants were set up by the year 1985-86 in different parts of he country. A large-scale urban waste recycling plant at Okhla (New Delhi) generates about 5 lakh cubic feet gas per day. In the VIIIth plan it is proposed that about 12 lakhs cubic feet gas will be generated per day. According to Water Sewage Disposal Undertaking 4,300 gas connections have been given in Delhi.

Conservation of energy

The need for conservation of energy becomes essential in view of the following simple facts:

i. Most of the activities of present day large and growing population require massive energy inputs, thus per capital and total demands for energy are fast increasing.

ii. The fossil fuels we are not endowed with which much are not to last for long at the present rate of consumption. At the same time much talked about possible alternate source of energy are, in fact, still in tentative state of development so far their practical applicability is concerned.

iii. There are economic constraints for many countries even if they want to develop conventional sources of energy, what to talk about sources like nuclear energy.

iv. Unplanned exploitation and use of energy result in widespread environmental damage, the cost of which is to be borne by humanity.

Conservation of energy means not simply the curtailment of energy consumption but to curtail unnecessary or wasteful consumption of energy and make thrifty use of energy producting resources to ensure their steady supply for generations.

7 The Wildlife (Protection) Act, 1972

An Act to provide for the protection of wild animals and birds and for matters connected therewith or ancillary or incidental thereto.

Whereas it is expedient to provide for the protection of wild animals and birds and for matters connected therewith or ancillary or incidental thereto;

And whereas Parliament has no power to make laws for the States with respect to any of the matters aforesaid as provided in Arts. 249 and 250 of the Constitution;

And whereas in pursuance of Cl. (1) of Art. 252 of the Constitution resolutions have been passed by all the Houses of the Legislatures of the States of Andhra Pradesh, Bihar, Gujarat, Haryana, Himachal Pradesh, Madhya Pradesh, Manipur, Punjab, Rajasthan, Uttar Pradesh and West Bengal to the effect that the matters aforesaid should be regulated in those States by Parliament by law;

Be it enacted by Parliament in the Twenty-third Year of Republic of India as follows:

CHAPTER I
Preliminary

1. Short title, extent and commencement

(1) This Act may be called the Wild Life (Protection) Act, 1972.

(2) It extends, in the first instance, to the whole of the States of Andhra Pradesh, Bihar, Gujarat, Haryana, Himachal Pradesh, Madhya Pradesh, Manipur, Punjab, Rajasthan, Uttar Pradesh and West Bengal and to all the Union territories; and it shall also extend to such other State as may adopt this Act by resolution passed in that behalf in pursuance of Cl. (1) of Art. 252 of the Constitution.

(3) It shall come into force in a State of Union territory to which it extends, or may become extended in future, on such date as the Central Government may, by notification, appoint, and different dates may be appointed for different provisions of this Act or for different States or Union territories.

2. Definitions

In this Act, unless the context otherwise requires,—

(1) "animal" includes amphibians, birds, mammals and reptiles and their young, and also includes, in the cases of birds and reptiles, their eggs;

(2) "animal article" means an article made from any captive animal or wild animal, other than vermin, and includes an article or objective in which the whole or any part of such animals has been used;

(3) "big game" means any animal specified in Sch. III;

(4) "Board" means the Wild Life Advisory Board constituted under sub-section (1) of Sec. 6;

(5) "Captive animal" means any animal, specified in Sch. I, Sch. II, Sch. III or Sch. IV, which is captured or kept or bred in captivity;

(6) "cattle" includes buffaloes, bulls, bullocks, camels, cows, domestic elephants, donkeys, goats, horses, mules, pigs and sheep, and also includes their young:

(7) "Chief Wild Life Warden" means the person appointed as such under Cl. (a) of sub-section (1) of Sec. 4;

(8) "closed area" means the area which is declared under sub-section (1) of Sec. 37 to be closed to hunting;

(9) "Collector" means the chief officer in charge of the revenue administration of a district;

(10) "commencement of this Act", in relation to—

(a) a State, means commencement of this Act in that State,

(b) any provision of this Act, means the commencement of that provision in the concerned State;

(11) "dealer" means any person who carries on the business of buying and selling any captive animal, animal article, trophy, uncured trophy or meat;

(12) "Director" means the person appointed as Director of Wild Life Preservation under Cl. (a) of sub-section (1) of Sec. 3;

(13) "game reserve" means an area declared under Sec. 36 to be a game reserve;

(14) "Government property" means any property referred to in Sec. 39;

(15) "habitat" includes land, water or vegetation which is the natural home of any wild animal;

(16) "hunting", with its grammatical variations and cognate expressions, includes,—

(a) capturing, killing, poisoning, snaring and trapping of any wild animal and every attempt to do so,

(b) driving any wild animal for any of the purposes specified in the sub-clause (a),

(c) injuring or destroying or taking any part of the body of any such animal or, in the case of wild birds or reptiles, damaging the eggs of such birds or reptiles, or disturbing the eggs or nests of such birds or reptiles;

(17) "land" includes canals, creeks and other water channels, reservoirs, rivers, streams and lakes, whether artificial or natural, and also includes boulders and rocks;

(18) "licence" means a licence granted under this Act;

(19) "manufacturer" means a manufacturer of animal articles;

(20) "meat" includes blood, bones, sinew, eggs, fat and flesh, whether raw or cooked, of nay wild animal, other than vermin;

(21) "National Park" means an area declared, whether under Sec. 35 or Sec. 38, or deemed, under sub-section (3) of Sec. 66, to be declared, as a National Park;

(22) "notification" means a notification published in the official Gazette;

(23) "permit" means a permit granted under this Act or any rule made thereunder;

(24) "person" includes a firm;

(25) "prescribed" means prescribed by rules made under this Act;

(26) "sanctuary" means an area declared, whether under Sec. 18 or Sec. 38, or deemed, under sub-section (3) of Sec. 66, to be declared, as a wild life sanctuary;

(27) "small game" means any animal specified in Sch. IV;

(28) "special game" means any animal specified in Sch. II;

(29) "State Government" in relation to a Union territory, means the Administrator of that Union territory appointed by the President under Art. 239 of the Constitution;

(30) "taxidermy", with its grammatical variations and cognate expressions, means the curing, preparation or preservation of trophies;

(31) "trophy" means the whole or any part of any captive animal or wild animal, other than vermin, which has been kept or preserved by any means, whether artificial or natural, and includes—

(a) rugs, skins and specimens of such animal mounted in whole or in part through a process of taxidermy, and

(b) antler, horn, rhinoceros horn, hair, feather, nail, tooth, musk, eggs, and nests;

(32) "uncured trophy" means the whole or any part of any captive animal or wild animal, other than vermin, which has not undergone a process of taxidermy, and includes a freshly killed wild animal;

(33) "vehicle" means any conveyance used for movement on land, water or air and includes buffalo, bull, bullock, camel, donkey, elephant, horse and mule;

(34) "vermin" means any wild animal specified in Sch. V;

(35) "weapon" includes ammunition, bows and arrows, explosives, fire-arms, hooks, knives, nets, poison, snares and traps and any instrument or apparatus capable of anesthetizing, decoying, destroying injuring or killing an animal;

(36) "wild animal" means any animal found wild in nature and includes any animal specified in Sch I, Sch. II, Sch. III, Sch. IV or Sch. V, wherever found;

(37) "wild life" includes any animal, bees, butterflies, crustacea, fish and moths; and aquatic or land vegetation which forms part of any habitat;

(38) "Wile Life Warden" means the person appointed as such under Cl. (b) for sub-section (1) of Sec. 4.

CHAPTER II

Authorities to be Appointed or Constituted Under The Act

3. Appointment of Director and other officers

(1) The Central Government may, for the purposes of this Act, appoint, -

(a) A Director of Wild Life Preservation;

(b) Assistant Directors of Wild Life Preservation; and

(c) Such other officers and employees as may be necessary.

(2) In the performance of his duties and exercise of his powers by or under this Act, the Director shall be subject to such general or special directions, as the Central Government may, from time to time, give.

(3) The Assistant Directors of Wild Life Preservation and other officers and employees appointed under this section shall be subordinate to the Director.

4. Appointment of Chief Wild Life Warden and other officers

(1) The State Government may, for the purposes of this Act, appoint -

(a) a Chief Wild Life Warden;

(b) Wild Life Wardens; and

(c) such other officers and employees as may be necessary.

(2) In the performance of his duties and exercise of his powers by or under this Act, the Chief Wild Life Warden shall be subject to such general or special directions, as the State Government may, from time to time, give.

(3) The Wild Life Warden and other officers and employees appointed under this section shall be subordinate to the Chief Wild Life Warden.

5. Power to delegate

(1) The Director may, with the previous approval of the Central Government, by order in writing delegate all or any of his powers and duties under this Act to any officer subordinate to him subject to such condition, if any, as may be specified in the order.

(2) The Chief Wild Life Warden may, with the previous approval of the State Government, by order in writing, delegate all or any of his powers and duties under this Act, except those under Cl. (a) of sub-section (I) of Sec. 11, to any officer subordinate to his subject to such conditions, if any, as may be specified in the order.

(3) Subject to any general or special direction given or condition imposed by the Director or the Chief Wild Life Warden, any person, authorised by the Director or the Chief Wild Life Warden to exercise any powers, may exercise those powers in the same manner and to the same effect as if they had been conferred on that person directly by this Act and not by way of delegation.

6. Constitution of Wild Life Advisory Board

(1) The State Government, or, in the case of a Union territory, the Administrator, shall, as soon as may be after the commencement of this Act, constitute a Wild Life Advisory Board consisting of the following members, namely:

(a) the Minister in charge of forests in the State or Union territory, or, if there is no such Minister, the Chief Secretary to the State Government, or, as the case may be, the Chief Secretary to the Government or the Union territory,who shall be the Chairman;

(b) two members of the State Legislature or, in the case of a Union territory having a Legislature, two members of the Legislature of the Union territory, as the case may be;

(c) Secretary to the State Government, or the Government of the Union territory, in charge of forests;

(d) Chief Conservator of Forests, *ex officio;*

(e) an officer to be nominated by the Director;

(f) Chief Wild Life Warden, *ex officio;*

(g) such other officers and non-officials, not exceeding fifteen, who, in the opinion of the State Government, are interested in the protection of wild life.

(2) The State Government shall appoint Chief Conservator of Forests or Chief Wild Life Warden as the Secretary of the Board.

(3) The term of office of the members of the Board referred to in Cl. (g) of sub-section (1) and the manner of filling vacancies among them shall be such as may be prescribed.

(4) The members shall be entitled to receive such allowances in respect of expenses incurred in the performance of their duties as the State Government may prescribe.

7. Procedure to be followed by the Board

(1) The Board shall meet at least twice a year at such place as the State Government may direct.

(2) The Board shall regulate its own procedure (including the quorum).

(3) No act or proceeding of the Board shall be invalid merely by reason of the existence of any vacancy there in or any defect in the constitution thereof or any irregularity in the procedure of the Board affecting the merits of the case.

8. Duties of Wild Life Advisory Board

It shall be the duty of the Wild Life Advisory Board to advise the State Government. -

(a) in the selection of areas to be declared as sanctuaries, National Parks, game reserves and closed areas and the administration thereof;

(b) in formulation of the policy in granting licences and permits under this Act;

(c) in any matter relating to any Schedule; and

(d) which may be referred to it by the State Government.

CHAPTER III
Huntings of Wild Animals

9. Hunting of wild animals

(1) No person shall hunt any wild animal specified in Sch. I.

(2) No person shall hunt any wild animal specified in Sch. II, Sch. III, or Sch. IV, except under, and in accordance with, the conditions specified in a licence granted under sub-section (5).

(3) Any person desiring to obtain a licence referred to in sub-section (2) shall apply, in such form, and on payment of such fee, as may be prescribed, to the Chief Wild Life Warden or any other officer authorised by the State government in this behalf (hereinafter referred to as the authorised officer).

(4) An application under sub-section (3) may be made for all or any of the following kinds of licences, namely;

(a) Special games hunting licence.

(b) Big game hunting licence.

(c) Small game hunting licence.

(d) Wild animal trapping licence.

Provided that any such licence may be restricted to the hunting or trapping of such animal as may be specified therein.

(5) On receipt of an application under sub-section (3),the Chief Wild Life Warden or the authorised officer may, after making such inquiry as he may think fit with regard to the fitness or otherwise of the applicant to receive the licence, grant or refuse to grant such licence after recording in writing his reasons for such grant or refusal, and when the grant of a licence is refused, the fee paid therefore shall be refunded to the applicant.

(6) A licence granted under this section shall

(a) be in such form as may be prescribed;

(b) be valid for such period as may be specified therein;

(c) be subject to such terms, conditions and restrictions as may be prescribed;

(d) not be transferable.

10. Maintenance of records of wild animals killed or captured

(1) The holder of every licence shall maintain a record, containing such particulars as may be prescribed, of the wild animals, other than vermin, killed or captured by him during the currency of the licence.

(2) When any animal specified in Sch. II or Sch. III is killed, wounded, or captured by the holder of a licence, he shall, not latter than fifteen days from the date of such killing, wounding or capture, or before leaving the area specified in the licence, whichever is earlier, intimate in writing to the Chief Wild Life Warden or the authorised officer, the prescribed particulars of such animal killed, wounded or captured by him.

(3) The holder of every licence shall, not later than fifteen days from the date of expiry of such licence, surrender the records maintained by him under sub-section (1) and the licence to the Chief Wild Life Warden or the authorised officer, and shall sign a declaration, in the prescribed form certifying the accuracy of the records maintained by him under sub-section (1).

11. Hunting of wild animals to be permitted in certain cases

(1) Notwithstanding anything contained in any other law for the time being in force and subject to the provision of Chapter IV,

(a) the Chief Wild Life Warden may, if he is satisfied that any wild animal specified in Sch. I has become dangerous to human life or is so disabled or diseased as to be beyond recovery, by order in writing and stating the reasons there for, permit any person to hunt such animal or cause such animal to be hunted;

(b) the Chief Wild Life Warden or the authorised officer may, if he is satisfied that any wild animal specified in Sch. II, Sch. III, or Sch. IV, has become dangerous to human life or to property (including standing crops on any land) or is so disabled or diseased as to be beyond recovery, by order in writing and stating the reasons therefore, permit any person to hunt such animal or cause such animal to be hunted.

(2) The killing or wounding in good faith of any wild animal in defence on oneself or of any other person shall not be an offence:

Provided that nothing in this sub-section shall exonerate any person who, when such defence becomes necessary, was committing any act in contravention of any provision of this Act or any rule or order made thereunder.

(3) Any wild animal killed or wounded in defence of any person shall be Government property.

12. Grant of permit for special purposes

Notwithstanding anything contained else where in this Act, it shall be lawful for the Chief Wild Life Warden, to grant, a permit, by an order in writing stating the reasons there for, to any person, on payment of such fee as may be prescribed, which shall entitle the holder of such permit to hunt, subject to such conditions as may be specified therein, any wild animal specified in such permit, for the purpose of, -

(a) education;

(b) scientific research;

(bb) scientific management;

Explanation. - For the purposes of Cl. (bb), the expression "scientific management" means -

(i) translocation of any wild animals to an alternative suitable habitat; or

(ii) population management of wild life, without killing or poisoning or destroying any wild animals; or

(c) collection of specimens for zoological gardens, museums and similar institutions:Provided that no such permit shall be granted

(a) in respect of any wild animal specified in Sch. I, except with the previous permission of the Central Government, and

(b) in respect of any other wild animal, except with the previous permission of the State Government.

13. Suspension or cancellation of licence

The Chief Wild Life Warden or the authorised officer may, subject to any general or special order of the State Government, for good and sufficient reasons, to be recorded in writing, suspend or cancel any licence granted under this Chapter:

Provided that no such suspension or cancellation shall be made except after giving the holder of the licence a reasonable opportunity of being heard.

14. Appeals

(1) An appeal from an order refusing to grant a licence under Sec. 9, or an order suspending or cancelling a licence under Sec. 13, shall lie, -

(a) if the order is made by the authorised officer, to the Chief Wild Life Warden, or

(b) if the order is made by the Chief Wild Life Warden, to the State Government.

(2) In the case of an order passed in appeal by the Chief Wild Life Warden under sub-section (1), a second appeal shall lie to the Sate Government.

(3) subject as aforesaid, every order passed in appeal under this section shall be final.

(4) No appeal shall be entertained unless it is preferred within fifteen days from the date of communication to the applicant of the order appealed against:

Provided that the appellate authority may admit any appeal after the expiry of the period aforesaid, if it is satisfied that the appellant had sufficient cause for not preferring the appeal in time.

15. Hunting of young and female of wild animals

No person shall, unless specially authorised by a licence, hunt the young of any wild animal, other than vermin, or any female of such animal, or any deer with antlers in velvet.

16. Declaration of closed time

(1) The State Government may, by notification, declare the whole year or any part thereof, to be a closed time throughout the State, or any part thereof, for such wild animal as may be specified in the notification and to hunting of such animal shall be permitted, during the said period, in the area specified in the notification.

(2) The provisions of sub-section (I) shall not apply to vermin unless otherwise specified by the State Government in this behalf.

17. Restrictions of hunting

(1) No person shall -

(a) hunt any wild animal, from or by means of, a wheeled or a mechanically propelled vehicle on water or land, or by aircraft;

(b) use an aircraft, motor vehicle or launch for the purpose of driving or stampeding any wild animal;

(c) hunt any wild animal with chemicals, explosives, nets, pitfalls, poisons, poisoned-weapons, snares or traps, except in so far as they relate to the capture of wild animals under a Wild Animal Trapping Licence;

(d) hunt any special game or big game other than with a rifle unless specially authorised by the licence to hunt with a shot-gun using single-slug bullets;

(e) for the purpose of hunting, set fire to any vegetation;

(f) use any artificial light for the purpose of hunting, except when specially authorised to do so under licence in the case of carnivora over a kill;

(g) hunt any wild animal during the hours of night,that, that is to say, between sunset and sunrise, except when specially authorised to do so under a licence in the case of carnivora over a kill;

(h) hunt any wild animal on a salt-lick or water hole or other drinking place or on path or approach to the same, except sand-grouse and water-birds;

(i) hunt any wild animal on any land not owned by Government, without the consent of the owner or his agent or the lawful occupier of such land;

(j) notwithstanding that he holds a licence for the purpose, hunt any wild animal during the closed time referred to in Sec. 16;

(k) hunt, with the help of dogs, any wild animal except water-bird, chakor, partridge or quail.

(2) The provisions of sub-section (1) shall not apply to vermin.

CHAPTER IV

Sanctuaries, National Parks, Game Reserves and Closed Areas

18. Declaration of sanctuary

(1) The State Government may, by notification, declare any area to be a sanctuary if it considers that such area is of adequate ecological, faunal, floral, geomorphological, natural or zoological significance, for the purpose of protecting, propagating wild life or its environment.

(2) The notification referred to in sub-section (1) shall specify, as nearly as possible, the situation and limits of such area.

Explanation.—For the purposes of this section, it shall be sufficient to describe the area by roads, rivers, ridges or other well-known or readily intelligible boundaries.

19. Collector to determine rights

Whenever any area is decalred to be a sanctuary, the Collector shall inquire into, and determine the existence, nature and extent of the rights of any person in or over the land comprised within the limits of the sanctuary.

20. Bar of accrual or rights

After the issue of a notification under Sec. 18, no right shall be acquired in, on or over the land comprised within the limits of the area specified in such notification, except by succession, testamentary or intestate.

21. Proclamation by Collector

When a notification has been issued under Sec. 18, the Collector shall publish in the regional languages in every town and village in or in the neighbourhood of the area comprised therein, a proclamation -

(a) specifying, as nearly as possible, the situation and the limits of the sanctuary; and

(b) requiring any person, claiming any right mentioned in Sec. 19, to prefer before the Collector, within two months from the date of such proclamation, a written claim in the prescribed form specifying the nature and extent of such right with necessary details and the amount and particulars of the compensation, if any, claimed in respect thereof.

22. Inquiry by Collector

The Collector shall, after service of the prescribed notice upon the claimant, expeditiously inquire into-

(a) the claim preferred before him under Cl. (b) of Sec. 21, and

(b) the existence of any right mentioned in Sec. 19 and not claimed under Cl. (b) of Sec. 21, so far as the same may be ascertainable from the records of the State government and the evidence of any person acquainted with the same.

23. Powers of Collector

For the purpose of such inquiry, the Collector may exercise the following powers, namely;

(a) the power to enter in or upon any land and to survey, demarcate and make a map of the same or to authorise any other officer to do so;

(b) the same powers as are vested in a civil court for the trial of suits.

24. Acquisition of rights

(1) In the case of a claim to a right in or over any land referred to in Sec. 19, the Collector shall pass an order admitting or rejecting the same in whole or in part.

(2) If such claim is admitted in whole or in part, the Collector may either -

(a) exclude such land from the limits of the proposes sanctuary, or

(b) proceed to acquire such land or rights, except where by an agreement between the owner of such land or holder of rights and the Government the owner or holder of such rights has agreed to surrender his rights to the Government, in or over such land, and on payment of such compensation, as is provided in the Land Acquisition Act, 1894 (1 of 1894).

25. Acquisition proceedings

(1) For the purpose of acquiring such land, or rights in or over such land, -

(a) the Collector shall be deemed to be a Collector, proceeding under the Land Acquisition Act, 1894 (1 of 1894);

(b) the claimant shall be deemed to be a person interested and appearing before him in pursuance of a notice given under Sec.9 of that Act;

(c) the provisions of the sections, preceding Sec.9 of that Act, shall be deemed to have been complied with;

(d) where the claimant does not accept the award made in his favour in the matter of compensation, he shall be deemed, within the meaning of Sec. 18 of that Act, to be a person interested who has not accepted the award, and shall be entitled to proceed to claim relief against the award under the provision of Part III of that Act;

(e) the Collector, with the consent of the claimant, or the Court, with the consent of both the parties, may award compensation in land or money or partly in land and partly in money; and

(f) in the case of the stoppage of a public way or a common pasture, the Collector may, with the previous sanction of the State Government, provide for an alternative public way or common pasture, as far as may be practicable or convenient.

(2) The acquisition under this Act of any land or interest therein shall be deemed to be acquisition for a public purpose.

26. Delegation of Collector's powers

The State Government may, by general or special order

direct that the powers exercisable or the functions to be performed by the Collector under Secs. 19 to 25 (both inclusive) may be exercised and performed by such other officer as may be specified in the order.

27. Restriction on entry in sanctuary

(1) No person other than,—

(a) a public servant on duty,

(b) a person who has been permitted by the Chief Wild Life Warden or the authorised officer to reside within the limits of the sanctuary,

(c) a person who has any right over immovable property within the limits of the sanctuary,

(d) a person passing through the sanctuary along a public highway, and

(e) the dependants of the person referred to in Cl.(a), Cl. (b) or Cl. (c) shall enter or reside in the sanctuary, except under and in accordance with the conditions of a permit granted under Sec.28.

(2) Every person shall, so long as he resides in the sanctuary, be bound—

(a) to prevent the commission, in the sanctuary, of an offence against this Act;

(b) where there is reason to believe that any such offence against this Act has been committed in such sanctuary, to help in discovering and arresting the offender;

(c) to report the death of any wild animal and to safeguard its remains until the Chief Wild Life Warden or the authorised officer takes charge thereof;

(d) to extinguish any fire in such sanctuary of which he has knowledge or information and to prevent from spreading, by any lawful means in his power, any fire within the vicinity of such sanctuary of which he has knowledge or information; and

(e) to assist any forest officer, Chief Wild Life Warden, Wild Life Warden or police officer demanding his aid for preventing the commission of any offence against this Act or in the investigation of any such offence.

28. Grant of permit

(1) The Chief Wild Life Warden may, on application, grant to any person a permit to enter or reside in a sanctuary for all or any of the following purposes, namely:

(a) investigation or study of wild life and purposes ancillary or incidental thereto;

(b) photography;

(c) scientific research;

(d) tourism;

(e) transaction of lawful business with any person residing in the sanctuary.

(2) A permit to enter or reside in a sanctuary shall be issued subject to such conditions and on payment of such fee as may be prescribed.

29. Hunting in sanctuary without permit prohibited

(1) Notwithstanding anything contained elsewhere in this Act, no person shall hunt any wild animal in a sanctuary or remove therefrom any wild animal, whether alive or dead, or any trophy, uncured trophy, or meat derived from such animal:

Provided that if the Chief Wild Life Warden is satisfied that it is necessary that any wild animal in a sanctuary should be hunted or removed,-

(a) for the better protection of wild life, or

(b) for any other good and sufficient reason,

he may, with the previous approval of the State Government, grant a permit authorising any reason to hunt or remove such wild animal under the direction of an officer authorised by him or cause it to be hunted or removed.

(2) A permit granted under sub-section (1) shall specify the kind and number of wild animals that may be hunted or removed by the holder of such permit.

(3) The Chief Wild Life Warden may,for good and sufficient reason, to be recorded in writing cancel any permit granted under Sec. 28 or under this section:

Provided that no such cancellation shall be made except after giving the holder of the permit a reasonable opportunity of being heard.

(4) Any person aggrieved by the cancellation of a permit under sub-section (3) may, within fifteen days from the date of such cancellation appeal to the State Government, whose decision shall be final:

Provided that the State Government may admit any appeal preferred after the expiry of the period aforesaid if it is satisfied that the appellant had sufficient cause for not preferring the appeal in time.

30. Causing fire prohibited

No person shall set fire to a sanctuary, or kindle any fire, or leave any fire burning, in a sanctuary, in such manner as to endanger such sanctuary.

31. Prohibition of entry into sanctuary with weapon

No person shall enter a sanctuary with any weapon except with the previous permission in writing of Chief Wild Life Warden or the authorised officer.

32. Ban on use of injurious substances

No person shall use, in a sanctuary, chemicals, explosive or any other substances which may cause injury to, or endanger, any wild life in such sanctuary.

33. Control of sanctuaries

The Chief Wild Life Warden shall be the authority who shall control, manage and maintain all sanctuaries and for that purpose, within the limits of any sanctuary,—

(a) may construct such roads, bridges, buildings, fences or barrier gates, and carry out such other works as he may consider necessary for the purpose of such sanctuary;

(b) shall take such steps as will ensure the security of wild animals in the sanctuary and the preservation of the sanctuary and wild animals therein;

(c) may take such measures, in the interests of wild life, as he may consider necessary for the improvement of any habitat;

(d) may regulate, control or prohibit, in keeping with the interests of wild life, the grazing or movement of cattle;

(e) may regulate, control or prohibit, any fishing.

34. Registration of certain persons in possession of arms

(1) Within three months from the declaration of any area as a a sanctuary, every person residing in or within ten kilometres of any such sanctuary and holding a licence granted under the Arms Act, 1959

(54 of 1959), for the possession of arms or exempted from the provisions of that Act and possessing arms, shall apply in such form, on payment of such fee and within such time as may be prescribed, to the Chief Wild Life Warden or the authorised officer, for the registration of his name.

(2) On receipt of an application under sub-section (1), the Chief Wild Life Warden or the authorised officer shall register the name of the applicant in such manner as may be prescribed.

National Parks

35. Declaration of National Parks

(1) Whenever it appears to the State Government that an area, whether within a sanctuary or not, is, by reason of its ecological, faunal, floral, geomorphological or zoological association or importance, needed to be constituted as National Park for the purpose of protecting, propagating or developing wild life therein or its environment, it may, by notification, declare its intention to constitute such area as a National Park.

(2) The notification referred to in sub-section (1) shall define the limits of the area which is intended to be declared as a National Park.

(3) Where any area is intended to be declared as a National Park, the provisions of Secs. 19 to 26 (both inclusive) shall,as far as may be, apply to the investigation and determination of claims, and extinguishment of rights, in relation to any land in such area as they apply to the said matters in relation to any land in a sanctuary.

(4) When the following events have occured, namely,-

(a) the period for preferring claims has elapsed,

and all claims, if any, made in relation to any land in an area intended to be declared as a National Park, have been disposed of by the State Government, and

(b) all rights in respect of lands proposed to be included in the National Park have become vested in the State Government.

The State Government shall publish a notification specifying the limits of the area which shall be comprised within the National Park and declare that the said area shall be a National Park on and from such date as may be specified in the notification.

(5) No alteration of the boundaries of a National Park shall be made except on a resolution passed by the Legislature of the State.

(6) No person shall destroy, exploit or remove any wild life from a National Park or destroy or damage the habitat of any wild animal or deprive any wild animal of its habitat within such National Park except under and in accordance with a permit granted by the Chief Wild Life Warden and no such permit shall be granted unless the State Government, being satisfied that such destruction, exploitation or removal of wild life from the National Park is necessary for the improvement and better management of wild life therein, authorises the issue of such permit.

(7) No grazing of any cattle shall be permitted in a National Park and no cattle shall be allowed to enter therein except where such cattle is used as a vehicle by a person authorised to enter such National Park.

(8) The provisions of Secs.27 and 28, Secs. 30 to 32 (both inclusive), and Cls. (a),(b) and (c) of Sec.33 and Sec.34 shall, as far as may be apply, in relation to a

National Park as they apply in relation to a sanctuary.

Game Reserve

36. Declaration of game reserve

(1) The State Government may, by notification, declare any area to be a game reserve.

(2) No hunting of any wild animal shall be permitted in such reserve except under and in accordance with a licence issued under this section by the Chief Wild Life Warden or the authorised officer.

Closed Area

37. Declaration of closed area

(1) The State Government may,by notification, declare any area closed to hunting for such period as may be specified in the notification.

(2) No hunting of any wild animal shall be permitted in a closed area during the period specified in the notification referred to in sub-section (1).

Sanctuaries or National Parks declared by Central Government

38. Power of Central Government to declare areas as sanctuaries or National Park

(1) Where the State Government leases or otherwise transfers any area under its control, not being an area, within a sanctuary, to the Central Government, the Central Government may, if it is satisfied that the conditions specified in Sec.18 are fulfilled in relation to the area so transferred to it, declare such area, by notification, to be a sanctuary and the provisions of Secs. 19 to 35 (both inclusive), 54 and 55 shall apply in relation to such sanctuary as they apply in relation to a sanctuary declared by the State Government.

(2) The Central Government may, if it is satisfied that the conditions specified in Sec.35 are fulfilled in relation to any area referred to in sub-section (1), whether or not such area has been declared, to be a sanctuary by the Central Government or the State Government, declare such area, by notification, to be a National Park and the provisions of Secs. 35, 54 and 55 shall apply in relation to a National Park declared by the State Government.

(3) In relation to a sanctuary or National Park declared by the Central Government, the powers and duties of the Chief Wild Life Warden under the section referred to in sub-sections (1) and (2), shall be exercised and discharged by the Director or by such other officer as may be authorised by the Director in this behalf and references, in the sections aforesaid, to the State Government shall be constructed as references to the Central Government and reference therein to the Legislature of the State shall be constructed as a reference to Parliament.

CHAPTER V

Trade or Commerce in Wild Animals, Animal Article and Trophies

39. Wild animals, etc. to be Government property

(1) Every-

(a) wild animal, other than vermin, which is hunted under Sec. 11 or sub-section (1) of Sec.29 or sub-section (6) of Sec.35 or kept or bred in captivity in contravention of any provision of this Act or any rule or order made thereunder or found dead, or killed without a licence or by mistake; and

(b) animal article, trophy or uncured trophy or meat derived from any wild animal referred to in Cl. (a) in respect of which any offence against this Act or any rule or order made thereunder has been committed, shall be the property of the State Government, and, where such animal is hunted in a sanctuary of National Park declared by the Central Government, such animal or any article, trophy, uncured trophy or meat derived from such animal, shall be the property of the Central Government.

(2) Any person who obtains, by any means, the possession of Government property, shall, within forty-eight hours obtaining such possession to the nearest police station or the authorised officer and shall, if so required, hand over such property to the officer-in-charge of such police station or such authorised officer, as the case may be.

(3) No person shall, without the previous permission in writing of the Chief Wild Life Warden or the authorised officer -

(a) acquire or keep in his possession, custody or control, or

(b) transfer to any person, whether by way of gift, sale or otherwise, or

(c) destroy or damage, such Government property.

40. Declarations

(1) Every person having at the commencement of this Act the control, custody or possession of any captive animal specified in Sch. I or Part II of Sch. II, or any uncured trophy derived from such animal or salted or dried skins of such animal or the musk of a musk deer or the horn or a rhinoceros, shall, within thirty days from the commencement of this Act, declare to

the Chief Wild Life Warden or the authorised officer the number and description of the animal, or article of the foregoing description under his control, custody or possession and the place where such animal or article is kept.

(2) No person shall, after the commencement of this Act, acquire, receive, keep in his control, custody or possession, sell, offer for sale or otherwise transfer or transport any animal specified in Sch. I or Part II of Sch.II or any uncured trophy or meat derived from such animal, or the salted or dried skins of such animal or the musk of musk deer or the horn of a rhinoceros, except with the previous permission in writing of the Chief Wild Life Warden or the authorised officer.

(3) Nothing in sub-section (1) or sub-section (2) shall apply to a public museum or zoo.

(4) The State Government may, by notification, require any person to declare to the Chief Wild Life Warden or the authorised officer any animal, article or trophy (other than a musk of a musk deer or horn of a rhinoceros) or salted or dried skins, derived from an animal specified in Sch. I or Part II of Sch. II in his control, custody or possession in such form, in such manner, and within such time as may be prescribed.

41. Inquiry and preparation of inventories

(1) On receipt of a declaration made under Sec.40, the Chief Wild Life Warden or the authorised officer may, after such notice, in such manner and at such time, as may be prescribed,-

(a) enter upon the premises of a person referred to in Sec.40;

(b) make inquiries and prepare inventories of animal articles, trophies, uncured trophies, salted and dried skins and captive animals specified in Sch. I and Part II of Sch. II and found thereon; and

(c) affix upon the animals, animal articles, trophies or uncured trophies identification marks in such manner as may be prescribed.

(2) No person shall obliterate or counterfeit any identification mark referred to in this Chapter.

42. Certificate of ownership

The Chief Wild Life Warden may, for the purpose of Sec.40, issue a certificate of ownership in such form, as may be prescribed, to any person who, in his opinion, is in lawful possession of any wild animal or any animal article, trophy, uncured trophy and may, where possible, mark, in the prescribed manner, such animal article, trophy or uncured trophy for purpose of identification.

43. Regulation of transfer of animal, etc.

(1) Subject to the provisions of sub-sections (2), (3) and (4), a person (other than a dealer) who does not possess a certificate of ownership shall not-

(a) sell or offer for sale or transfer whether by way of sale, gift or otherwise, any wild animal specified in Sch. I or Part II of Sch. II or any captive animal belonging to that category or any animal article, trophy, uncured trophy, or meat derived therefrom;

(b) make animal articles containing part or whole of such animal;

(c) put under a process of taxidermy an uncured trophy of such animal, except with the previous permission in writing of the Chief Wild Life Warden or the authorised officer.

(2) Where a person transfers or transports from the State in which he resides to another State or acquires by transfer from outside the State any such animal, animal article, trophy or uncured trophy as referred to in sub-section (1) in respect of which he has a certificate of ownership, he shall, within thirty days of the transfer or transport, report the transfer or transport to the Chief Wild Life Warden or the authorised officer within whose jurisdiction the transfer or transport is effected.

(3) No person who does not possess a certificate of ownership shall transfer or transport from one State to another State or acquire by transfer from outside the State any such animal, animal article, trophy or uncured trophy as is referred to in sub-section (1) except with the previous permission in writing of the Chief Wild Life Warden or the authorised officer within whose jurisdiction the transfer is to be effected.

(4) Before granting any permission under sub-section (1) or sub-section (3), the Chief Wild Life Warden or the authorised officer shall satisfy himself that the animal or article referred to therein has been lawfully acquired.

(5) While permitting the transfer or transport of animal. animal article, trophy or uncured trophy, as is referred to in sub-section (1), the Chief Wild Life Warden or the authorised officer -

(a) shall issue a certificate of ownership after such inquiry as he may deem fit;

(b) shall, where the certificate of ownership existed in the name of the previous owner, issue a fresh certificate of ownership in the name of the person to whom the transfer has been effected.

(c) may affix an identification mark on any such animal,animal article. trophy or uncured trophy.

(6) Nothing in this section shall apply -

(a) to animal articles or trophies made out of feathers of peacocks;

(b) to any transaction entered into by a public museum or zoo with any other public museum or zoo.

44. Dealings in trophy and animal articles without licence prohibited

(1) Subject to the provisions of Chapter V-A, no person shall, except under, and in accordance with, a licence granted under sub-section (4),-

(a) commence or carry on the business as-

(i) a manufacturer of, or dealer in, any animal article; or

(i-a) a manufacturer of, or dealer in, any article made of ivory imported into India;

(ii) a taxidermist; or

(iii) a dealer in trophy or uncured trophy; or

(iv) a dealer in captive animal; or

(v) a dealer in meat; or

(b) cook or serve meat in any eating-house:

Provided that nothing in this sub-section shall prevent a person, who, immediately before the commencement of this Act was carrying on the business or occupation specified in this sub-section, from carrying on such business or occupation for a period of thirty days from such commencement, or where he has made an application within that period for the grant of a licence to him, until the licence is granted to him or he is informed in writing that a licence cannot be granted to him:

Provided further that nothing in this sub-section shall apply to manufacturers of, and dealers in, articles made of feathers of peacock.

Explanation For the purpose of this section, "eating-house" includes a hotel, restaurant or any other place where any eatable is served on payment, whether or not such payment is separately made for such eatable or is included in the amount charge for board and lodging.

(2) Every manufacturer of, or dealer in, animal article, or every dealer in captive animals. trophies or uncured trophies, or every taxidermist shall, within fifteen days from the commencement of this Act, declare to the Chief Wild Life Warden his stocks to animals articles, captive animals, trophies and uncured trophies, as the case may be, as on the date or such declaration and the Chief Wild Life Warden or the authorised officer may place an identification mark on every animal article, captive animal, trophy or uncured trophy, as the case may be.

(3) Every person referred to in sub-section (1) who intends to obtain a licence, shall, make an application to the Chief Wild Life Warden or the authorised officer for the grant of a licence.

(4) (a) Every application referred to in sub-section (3) shall be made in such form and on payment of such fee as may be prescribed, to the Chief Wild Life Warden or the authorised Officer.

(b) No licence referred to the sub-section (1) shall

be granted unless the Chief Wild Life Warden or the authorised officer having regard to the antecedents and previous experience of the applicant, the implication which the grant of such licence would have on the status of wild life and to such other matters as may be prescribed in this behalf and after making such inquiry in respect of those matters as he may think fit, is satisfied that the licence should be granted.

(5) Every licence granted under this section shall specify the premises in which and the conditions, if any, subject to which the licensee shall carry on his business.

(6) Every licence granted under this section shall-

(a) be valid for one year from the date of its grant;

(b) not be transferable; and

(c) be renewable for a period not exceeding one year at a time.

(7) No application for the renewal of a licence shall be rejected unless the holder of such licence has been given a reasonable opportunity of presenting his case and unless the Chief Wild Life Warden or the authorised officer is satisfied that-

(i) the application for such renewal has been made after the expiry of the period specified thereof, or

(ii) any statement made by the applicant at the time of the grant or renewal of the licence was incorrect or false in material particulars, or

(iii) the applicant has contravened any term or condition of the licence or any provision of this Act or any rule made thereunder, or

(iv) the applicant does not fulfil the prescribed conditions.

(8) Every order granting or rejecting an application for the grant or renewal of a licence shall be made in writing.

(9) Nothing in the foregoing sub-sections shall apply in relation to vermin.

45. Suspension or cancellation of licences

Subject to any general or special order of the State Government, the Chief Wild Life Warden or the authorised officer may, for reasons to be recorded by him in writing, suspend or cancel any licence granted or renewed under Sec.44:

Provided that no such suspension or cancellation shall be made except after giving the holder of the licence a reasonable opportunity of being heard.

46. Purchase

(1) An appeal from an order refusing to grant or renew a licence under Sec.44 or an order suspending or cancelling a licence under Sec.45 shall lie -

(a) if the order is made by the authorised officer, to the Chief Wild Life Warden; or

(b) if the order is made by the Chief Wild Life Warden, to the State Government.

(2) In the case of an order passed in appeal by the Chief Wild Life Warden under Cl. (a) of sub-section (1), a second appeal shall lie to the State Government.

(3) Subject as aforesaid, every order passed in appeal under this section shall be final.

(4) An appeal under this section shall bepreferred within thirty days from the date of communication, to the applicant, of the order appealed against:

Provided that the appellate authority may admit any appeal preferred after the expiry of the period aforesaid if it is satisfied that the appellant had sufficient cause for not preferring the appeal in time.

47. Maintenance of records

A license under this Chapter shall

(a) keep records, and submit such returns of his dealings, as may be prescribed,

(i) to the Director or any other officer authorised by him in this behalf, and

(ii) to the Chief Wild Life Warden or the authorised officer; and

(b) make such records available on demand for inspection by such officers.

48. Purchase of animal, etc. by license

No licensee under this Chapter shall -

(a) keep in his control, custody or possession,-

(i) any animal, animal article, trophy or uncured trophy in respect of which a declaration under the provisions of sub-section (2) of Sec.44 has to be made but has not been made;

(ii) any animal or animal article, trophy, uncured trophy or meat which has not been lawfully acquired under the provisions of this Act or any rule or order made thereunder.

(b) (i) capture any wild animal, or

(ii) acquire, receive, keep in his control, custody or possession, or sell, offer for sale or transport, any captive animal specified in Sch. I or Part II of Sch. II or any animal article, trophy, uncured trophy or meat derived therefrom or serve such meat, or put under a process of taxidermy or make animal article containing part or whole of such animal, except in accordance with such rules as may be made under this act:

Provided that where the acquisition or, possession, control or custody of such animal or animal article, trophy or uncured trophy entails the transfer or transport from one State to another, no such transfer or transport shall be effected except with the previous permission in writing of the Director or any other officer authorised by him in this behalf:

Provided further that no such permission under the foregoing provison shall be granted unless the Director or the officer authorised by him is satisfied that the animal or article aforesaid has been lawfully acquired.

49. Purchase of captive animal, etc. by a person other than a licensee

No person shall purchase, receive or acquire any captive animal, wild animal, other than vermin, or any animal article, trophy, uncured trophy or meat derived therefrom otherwise than from a dealer or from a person authorised to sell or otherwise transfer the same under this Act:

Provided that nothing in this section shall apply to any transaction entered into by a public museum or zoo with any other public museum or zoo.

CHAPTER V-A

Prohibition of Trade or Commerce in Trophies, Animal Articles, Etc Derived From Certain Animals

49-A. Definitions

In this Chapter,-

(a) "schedules animal" means an animal specified for the time being in Sch.I or Part II of Sch.II;

(b) "scheduled animal article" means an article made from any scheduled animal and includes an article or object in which the whole or any part of such animal has been used;

(c) "specified date" means -

(i) in relation to a scheduled animal on the commencement of the Wild Life (Protection) Amendment Act, 1986, the date of expiry of two months from such commencement; and

(ii) in relation to any animal added or transferred to Sch. I or Part II of Sch. II at any time after such commencement, the date of expiry of two months from such addition or transfer.

49-B. Prohibition of dealings in trophies, animal articles, etc. derived from scheduled animals

(1) Subject to the other provisions of this section, on and after the specified date, no person shall,-

(a) commence or carry on the business as -

(i) a manufacturer of, or dealer in, scheduled animal articles; or

(ii) a taxidermist with respect to any scheduled animals of any parts of such animals; or

(iii) a dealer in trophy or uncured trophy derived from any scheduled animal; or

(iv) a dealer in any captive animals being scheduled animals; or

(v) a dealer in meat derived from any scheduled animal; or

(b) cook or serve meat derived from any scheduled animal in any eating-house.

Explanation- For the purpose of this sub-section, "eating-house" has the same meaning as in the Explanation below sub-section (1) of Sec.44.

(2) Subject to the other provisions of this section, no licence granted or renewed under Sec.44 before the specified date shall entitle the holder thereof or any other person to commence or carry on the business referred to in Cl. (a) of sub-section (1) of this section or the occupation referred to in Cl. (b) of that sub-section after such date.

(3) Notwithstanding anything contained in sub-section (1) or sub-section (2), where the Central Government is satisfied that is necessary or expedient so to do in the public interest, it may, by general or special order published in the Official Gazette, exempt, for purpose of export, any corporation owned or controlled by the Central Government (including a Government company within the meaning of Sec.617 of the Companies Act, 1956 (1 of 1956), or any society registered under the Societies Registration Act, 1860 (21 of 1860) or any other law for the time being in force, wholly or substantially financed by the Central Government, from the provisions of sub-sections (1) and (2).

(4) Notwithstanding anything contained in sub-section (1) or sub-section(2), but subject to any rules which may be made in this behalf, a person holding a licence under Sec.44 to carry on the business as a taxidermist may put under a process of taxidermy any scheduled animal or any part thereof.-

(a) for or on behalf of the Government or any corporation or society exempted under sub-section (3), or

(b) with the previous authorisation in writing of the Chief Wild Life Warden, for and on behalf of any person for educational or scientific purposes.

49-C. Declaration by dealers

(1) Every person carrying on the business or occupation referred to in sub-section (1) of Sec. 49-B shall, within thirty days from the specified date, declare to the Chief Wild Life Warden or the authorised officer,-

(a) his stocks, if any, as at the end of the specified date of-

(i) scheduled animal articles;

(ii) scheduled animals and parts thereof;

(iii) trophies and uncured trophies derived from scheduled animals;

(iv) captive animals, being scheduled animals;

(b) the place or places at which the stocks mentioned in the declaration are kept; and

(c) the description of such items, if any, of the stocks mentioned in the declaration which he desires to retain with himself for his bona fide personal use.

(2) On receipt of declaration under sub-section (1), the Chief Wild Life Warden or the authorised officer may take all or any of the measures specified in Sec. 41 and for this purpose, the provisions of Sec. 41 shall, so far as may be, apply.

(3) Where, in a declaration made under sub-section (1), the person making the declaration expresses his desire to remain with himself any of the items of the stocks specified in the declaration for the bona-fide personal use, the Chief Wild Life Warden, with the prior approval of the director, may, if he is satisfied

that the person is in lawful possession of such items, issue certificates of ownership in favour of such person with respect to all, or as the case may be, such of the items as in the opinion of the Chief Wild Life Warden, are required for the bonafide personal use of such person and affix upon such items identification marks in such manner as may be prescribed:

Provided that no such item shall be kept in any commercial premise.

(4) No person shall obliterate or counterfeit any identification mark referred to in sub-section (3).

(5, An appeal shall lie against any refusal to grant certificate of ownership under sub-section (3) and the provisions of sub-section(2),(3) and (4) of Sec.46 shall, so far as may be, apply in relation to appeals under this sub-section.

(6) Where a person who has been issued a certificate of ownership under sub-section (3) in respect of any item,-

(a) transfers such item to any person, whether by way of gift, sale or otherwise, or

(b) transfers or transports from the State in which he resides to another State any such item, he shall, within thirty days of such transfer or transport, report the transfer or transport to the Chief Wild Life Warden or the authorised officer within whose jurisdiction the transfer or transport is effected.

(7) No person, other than a person who has been issued a certificate or ownership under sub-section (3) shall, on and after the specified date, keep under his control, sell or offer for sale or transfer to any person any scheduled animal or a scheduled animal article.

CHAPTER VI

Prevention and Detection of offnces

50. Power of entry, search, arrest and detention

(1) Notwithstanding anything contained in any other law for the time being in force, the Director or any other officer authorised by him in this behalf or the Chief Wild Life Warden or the authorised officer or any forest officer or any police officer not below the rank of a sub-inspector, nay, if he has reasonable grounds for believing that any person has committed an offence against this Act,-

(a) require any such person to produce for inspection any captive animal, wild animal, animal article, meat, trophy or uncured trophy in his control, custody or possession, or any licence, permit or other document granted to him or required to be kept by him under the provisions of this Act;

(b) stop any vehicle or vessel in order to conduct search or inquiry or enter upon and search any premises, land, vehicle, or vessel, in the occupation of such person, and open and search any baggage or other things in his possession;

(c) seize any captive animal, wild animal, animal article, meat, trophy or uncured trophy in the possession of any person and appearing to him to be Government property, together with any trap, tool, vehicle, vessel or weapon used for committing any such offence, and unless he is satisfied that such person will appear and answer any charge which may be preferred against him, arrest him without warrant, and detain him.

(2) Any officer of a rank not inferior to that of an Assistant Director of Wild Life Preservation or Wild Life Warden, who or whose subordinate, has seized any trap, tool, vehicle or weapon under Cl. (c) of sub-section (1), may release the same, on the execution by the owner thereof of a bond for the production of the property so released, if and when so required, before the Magistrate having jurisdiction to try the offence an account of which the seizure has been made.

(3) It shall be lawful for any of the officers referred to in sub-section (1) to stop and detain any person, whom he sees doing any act for which a licence or permit is required under the provisions of this Act, for the purpose of requiring such person to produce the licence or permit and if such person fails to produce the licence or permit, as the case may be, he may be arrested without warrant, unless he furnished his name and address, and otherwise satisfies the officer arresting him that he will duly answer any summons or other proceedings which may be taken against him.

(4) Any person detained, or things seized under the foregoing power, shall forthwith be taken before a Magistrate to be dealt with according to law.

(5) Any person who,without reasonable cause, fails to produce anything, which he is required to produce under this section, shall be guilty of an offence against this Act.

(6) (a) Where any meat or uncured trophy is seized under the provisions of this section, the Assistant Director of Wild Life Preservation or any other officer of a gazetted rank authorised by him in this behalf or the chief Wild Life Warden or the authorised officer may arrange for the sale of the same and deal with the proceeds of such sale in such manner as may be prescribed.

(b) Where it is proved that the meat or uncured trophy is seized under the provisions of this section is not Government property, the proceeds of the sale shall be returned to the owner.

(7) Whenever any person is approached by any of the officers referred to in sub-section (1) for assistance in the prevention or detection of an offence against this Act, or in apprehending persons charged with the violation of this Act, or for seizure in accordance with Cl. (c) of sub-section (1), it shall be the duty of such person or persons to render such assistance.

51. Penalties

(1) Any person who contravenes any provision of this Act (except Chapter V-A) or any rule, order made thereunder or who commits a breach of any of the conditions of any licence or permit granted under this Act, shall be guilty of an offence against this Act, and shall on conviction, be punishable with imprisonment for a term which may extend to two years, or with fine which may extend to two thousand rupees, or with both:

Provided that where the offence committed is in relation to any animal specined in Sch. I or Part II of Sch. II or meat of any such animal or animal article, trophy or uncured trophy derived from such animal or where the offence relates to hunting in a sanctuary or a National Park, such offence shall be punishable with imprisonment for a term which shall not be less than six months but may extend to six years and also with fine which shall not be less than five hundred rupees:

Provided further that in the case of a second or subsequent offence of the nature mentioned in the foregoing proviso, the term of imprisonment shall not be less than one year and the amount of the fine shall not be less than one thousand rupees.

(1-A) any person who contravenes any provisions of Chapter V-A, shall be punishable with imprisonment for a term which shall not be less than one year but which may extend to seven years and also with fine which shall not be less than five thousand rupees.

(2) When any person is convicted of an offence against this Act, the Court trying the offence may order that any captive animal, wild animal, animal article, trophy, uncured trophy or meat in respect of which the offence has been committed and any trap, tool, vehicle, vessel, or weapon, used in the commission of the said offence be forfeited to the State Government and that any licence or permit, held by such person under the provisions of this Act, be cancelled.

(3) Such cancellation of licence or permit or such forfeiture shall be in addition to any other punishment that may be awarded for such offence.

(4) Where any person is convicted of an offence against this Act, the Court may direct that the licence, if any, granted to such person under, the Arms Act, 1959 (54 of 1959), for possession of any arm with which an offence against this Act has been committed, shall be cancelled and that such person shall not be eligible for a licence under the Arms Act, 1959, for a period of five years from the date of conviction.

52. Attempts and abetment

Whoever attempts to contravene, or abets the contravention of, any of the provisions of this Act or of any rule or order made thereunder shall be deemed to have contravened that provision or rule or order, as the case may be.

53. Punishment for wrongful seizure

If any person, exercising powers under this Act, vaxatiously and unnecessarily seizes the property of any other person on the pretence of seizing it for the reasons mentioned in Sec.50, he shall, on conviction, be punishable with imprisonment for a term which may extend to six months, or with fine which may extend to five hundred rupees, or with both.

54. Power to compound offences

(1) The State Government may, by notification, empower the Chief Wild Life Warden or any officer of a rank not inferior to that of a Deputy Conservator of Forests,-

(a) to accept, from any person against whom a reasonable suspicion exists that he has committed an offence against this Act, payment of a sum of money by way of composition of the offence which such person is suspected to have committed; and

(b) when any property has been seized is liable to be forfeited, to release the same on payment of the value thereof as estimated by such officer.

(2) On payment of such sum of money or such value, or both, as the case may be, to such officer, the suspected person, if in custody, shall be discharged, and the property, other than Government property, if any, seized, shall be released and no further proceedings in respect of the offence shall be taken against such person.

(3) The officer compounding any offence may order the cancellation of any licence or permit granted under this Act to the offender, or if not empowered to do so, may approach an officer so empowered, for the cancellation of such licence or permit.

(4) the sum of money accepted or agreed to be accepted as composition under Cl. (b) of sub-section (1) shall, in no case, exceed the sum of two thousand rupees:

Provided that no offence, for which a minimum period of imprisonment has been prescribed in sub-section (1) of Sec.51, shall be compounded.

55. Cognizance of offences

No Court shall take cognizance of any offence against this Act except on the complaint of the Chief Wild Life Warden or such other officer as the State Government may authorise in this behalf.

56. Operation of other laws not barred

Nothing in this Act shall be deemed to prevent any person from being prosecuted under any other law for the time being in force, for any act or omission which constitutes an offence against this Act or from being liable under such other law to any higher punishment or penalty than that provided by this Act:

Provided that no person shall be punished twice for the same offence.

57. Presumption to be made in certain cases

Where, in any prosecution for one offence against this Act, it is established that a person is in possession, custody or control of any captive animal, animal article, meat, trophy or uncured trophy, it shall be presumed, until the contrary is proved, the burden of proving which shall lie on the accused, that such person is in unlawful possession, custody or control of such captive animal, animal article, meat, trophy or uncured trophy.

58. Offences by companies

(1) Where an offence against this Act has been

committed by a company, every person who, at the time the offence was committed, was in charge of, and was responsible to, the company for the conduct of the business of the company as well as the company, shall be deemed to be guilty of the offence and shall be liable to be proceeded against and punished accordingly:

Provided that nothing contained in this sub-section shall render any such person liable to any punishment, if he proves that the offence was committed without his knowledge or that he exercised all due diligence to prevent the commission of such offence.

(2) notwithstanding anything contained in sub-section (1), where an offence against this Act has been committed by a company and it is proved that the offence has been committed with the consent or connivance of, or is attributable to any neglect on the part of, any director, manager, secretary or other officer of the company, such director, manager, secretary or other officer shall also be deemed to be guilty of that offence and shall be liable to be proceeded against and punished accordingly.

Explanation- For the purpose-of this section,-

(a) "company" means any body corporate and includes a firm or other association of individuals; and

(b) "director", in relation to a firm, means a partner in the firm.

CHAPTER VII

Miscellaneous

59. Officers to be public servants

Every officer referred to in Chapter II and every other

officer exercising any of the powers conferred by this Act shall be deemed to be a public servant within the meaning of Sec.21 of the Indian Penal Code (45 of 1860).

60 Protection of action taken in good faith

(1) No suit, prosecution or other legal proceeding shall lie against any officer or other employee of the Central Government or the State government for anything which is in good faith done or intended to be done under this Act.

(2) No suit or other legal proceeding shall lie against the Central Government or the State Government or any of its officers or other employees, for any damage caused or likely to be caused by anything which is in good faith done or intended to be done under this Act.

61. Power to alter entries in schedules

(1) The Central Government may, if it is of opinion that it is expedient so to do, by notification, add an entry to any Schedule or transfer any entry from one Part of a Schedule to another Part of the same Schedule or from one Schedule to another.

(2) The State Government may, if it is of opinion that it is expedient so to do, by notification, transfer any entry from Sch. II, Sch. III, Sch.IV or Sch. V to Sch. I and may also transfer any entry from Part I of Sch.II, or Sch.III, Sch.IV or Sch. V, to any other schedule.

(3) On the issue of a notification under sub-section (1) or sub-section (2), the relevant schedule shall be deemed to be altered accordingly, provided that every such alteration shall be without prejudice to anything done or omitted to be done before such alteration.

(4) If any alteration of any schedule made by the State Government under sub-section (2) is repugnant to any alteration made therein by the Central Government under sub-section (1), then, the alteration made by the Central Government, whether made before or after the notification made by the State Government, shall prevail and the alteration made by the State Government shall, to the extent of the repugnancy, be void:

Provided that any such alteration made by the State Government, if it has been made with the previous consent of the Central Government, shall prevail in that State:

Provided further that nothing in the foregoing proviso shall prevent the Central Government from modifying or cancelling, at any time, the alteration made by the State Government.

62. Declaration of certain wild animals to be vermin

Subject to the provisions of Sec.61, the State Government may, by notification, declare any wild animal other than those specified in Sch. I and Part II of Sch. II to be vermin for any area and for such period as may be specified therein and so long as such notification is in force, such wild animal shall be deemed to have been included in Sch. V.

63. Power of Central Government to make rules

(1) The Central Government may, by notification, make rules for all or any of the following, namely:

(a) the form in which declaration shall be made under sub-section (2) of Sec.44;

(aa) the matters to be prescribed under Cl. (b) of sub-section (4) of Sec.44;

(b) the terms and conditions which shall govern transactions referred to in Cl. (b) of Sec. 48;

(c) matter specified in sub-section (2) of Sec. 64 in so far as they relate to sanctuaries and National Parks declared by the Central Government;

(d) any other matter which is required to be, or may be prescribed.

(2) Every rule made under this section shall be laid, as soon as may be after it is made, before each House of Parliament, while it is in session, for a total period of thirty days which may be comprised in one session or in two or more successive sessions, and if before the expiry of the session immediately following the session or the successive sessions aforesaid, both Houses agree in making any modification in the rule or both Houses agree that the rule should not be made, the rule shall thereafter have effect only in such modified form or be or no effect, as the case may be; so however, that any such modification or annulment shall be without prejudice to the validity of anything previously done under that rule.

64. Power of State Government to make rules

(1) The State Government may, by notification, make rules for carrying out the provisions of this Act in respect of matters which do not fall within the purview of Sec.63.

(2) In particular and without prejudice to the generally of the foregoing power, such rules may provide for all or any of the following matters, namely:

(a) the term of office of the members of the Board referred to in Cl. (g) of sub-section (1) of Sec. 6 and the manner of filling vacancies among them;

(b) allowances referred to in sub-section (4) of Sec.6;

(c) the forms to be used for any application, certificate, claim, declaration, licence, permit, registration, return or other document, made, granted, or submitted under the provisions of this Act and the fees, if any, therefor;

(d) the conditions subject to which any licence or permit may be granted under this Act;

(e) the particulars of the record of wild animals, (captured or killed) to be kept and submitted by the licensee;

(f) regulation of the possession, transfer and the sale of captive animals, meat, animal articles, trophies and uncured trophies;

(g) regulation of taxidermy;

(h) any other matter which has to be, or may be, prescribed under this Act.

65. Rights of scheduled tribes to be protected

Nothing in this Act shall affect the hunting rights conferred on the Scheduled Tribes of the Nicobar Islands in the Union territory of Andaman and Nicobar Islands by notification of the Andaman and Nicobar Administration No.40/97/F. No. G-635, Vol. III, dated the 28th April, 1967, published at pages 1 to 5 of the Extraordinary issue of the Andaman and Nicobar Gazette, dated the 28th April, 1967.

66. Repeal and savings

(1) As from the commencement of this Act, every other Act relating to any matter contained in this Act and in force in a State shall, to the extent to which that Act or any provision contained therein corresponds,

or is repugnant, to this Act or any provision contained in this Act, stand repealed:

Provided that such repeal shall not,

(i) affect the previous operation of the Act so repealed, or anything duly done or suffered thereunder;

(ii) affect any right, privilege, obligation or liability acquired, accrued or incurred under the Act so repealed;

(iii) affect any penalty, forfeiture or punishment incurred in respect of any offence committed against the Act so repealed; or

(iv) affect any investigation, legal proceeding or remedy in respect of any such right, privilege, obligation, liability, penalty, forfeiture, or punishment as aforesaid; and any such penalty, forfeiture and punishment may be imposed, as if the aforesaid Act has not been repealed.

(2) Notwithstanding such repeal, -

(a) anything done or any action taken under the Act so repealed (including any notification, order, certificate, notice or receipt issued, application made, or permit granted) which is not inconsistent with the provisions of this Act be deemed to have been done or taken under the corresponding provisions of this Act as if this Act were in force at the time such thing was done or action was taken, and shall continue to be in force, unless and until superseded by anything done or any action taken under this Act;

(b) every licence granted under any Act so

repealed and in force immediately before the commencement of this Act shall be deemed to have been granted under the corresponding provisions of this Act and shall, subject to the provisions of this Act, continue to be in force for the unexpired portion of the period for which such licence had been granted.

(3) For the removal of doubts, it is hereby declared that any sanctuary or National Park declared by a State Government under any Act repealed under sub-section (1) shall be deemed to be a sanctuary or National Park, as the case may be, declared by the State Government under this Act and where any right in or over any land in any such National Park which had not been extinguished under the said Act, at or before the commencement of this Act, the extinguishment of such rights shall be made in accordance with the provisions of this Act.

INDEX